Cary Wolfe SERIES EDITOR

13 *Junkware*
Thierry Bardini

12 *A Foray into the Worlds of Animals and Humans, with* A Theory of Meaning
Jakob von Uexküll

11 *Insect Media: An Archaeology of Animals and Technology*
Jussi Parikka

10 *Cosmopolitics II*
Isabelle Stengers

9 *Cosmopolitics I*
Isabelle Stengers

8 *What Is Posthumanism?*
Cary Wolfe

7 *Political Affect: Connecting the Social and the Somatic*
John Protevi

6 *Animal Capital: Rendering Life in Biopolitical Times*
Nicole Shukin

5 *Dorsality: Thinking Back through Technology and Politics*
David Wills

4 *Bíos: Biopolitics and Philosophy*
Roberto Esposito

3 *When Species Meet*
Donna J. Haraway

2 *The Poetics of DNA*
Judith Roof

1 *The Parasite*
Michel Serres

JUNKWARE

Thierry Bardini

posthumanities 13

University of Minnesota Press
Minneapolis
London

Cover art by Vik Muniz: *Mound (California Rolls Headless Army Men Coke Caps Marijuana Poison Scorpions Feathers Wrist Watch Parts Desyrel Ô de Lancôme)*, *Mound (Rat Poison Jade Buddhas Pectin Drops Meteorites Taylor Pins Aspirin Granola Silver Glitter Killer Bees)*, *Mound (Dinosaur Dung Shredded Ezra Pound Canto [XIX] Dice AZT Plastic Babies Cat Hair Beetles)*, and *Mound (Brillo Pads Gold Scrap Viagra Jelly Beans Pubic Hair Curry Powder Metal Screws Edible Worms)*, all from the series *Mounds*, 2005. Courtesy of Sikkema, Jenkins and Co. gallery and Vik Muniz Studio. Art copyright Vik Muniz. Licensed by VAGA, New York, NY.

Chapter 1 was previously published as "How Junk Became Selfish: The Nominalist Breakdown of Molecular Biology," in *Biosemiotics Research Trends*, ed. Marcello Barbieri (Nova Science Publishers, 2007), 215–329; reprinted with permission from Nova Science Publishers, Inc. Chapter 6 was previously published as "Hypervirus: A Clinical Report," in *Critical Digital Studies: A Reader*, ed. Arthur Kroker and Marilouise Kroker (Toronto: University of Toronto Press, 2008), 143–57.

Lyrics to Pink Floyd's "Welcome to the Machine" appear courtesy of Roger Waters.

Translation of "The Eighth Elegy" in *Duino Elegies* (1922) by Rainer Maria Rilke appears courtesy of Alison Croggon.

Every effort was made to obtain permission to reproduce material in this book. If any proper acknowledgment has not been included here, we encourage copyright holders to notify the publisher.

Published by the University of Minnesota Press
111 Third Avenue South, Suite 290
Minneapolis, MN 55401-2520
http://www.upress.umn.edu

Library of Congress Cataloging-in-Publication Data

Bardini, Thierry.
Junkware / Thierry Bardini.
p. cm. — (Posthumanities ; v. 13)
Includes bibliographical references (p.) and index.
ISBN 978-0-8166-6750-5 (hc : alk. paper)
ISBN 978-0-8166-6751-2 (pb : alk. paper)
1. Philosophical anthropology. 2. DNA. I. Title.
BD450.B37 2010
128—dc22

2010003873

Printed in the United States of America on acid-free paper

The University of Minnesota is an equal-opportunity educator and employer.

18 17 16 15 14 13 12 11 10 9 8 7 6 5 4 3 2

To Fabienne and Léonard,
without whom my life would be only junk.

Junk is not, like alcohol or weed,
a means to increase enjoyment of life.
Junk is not a kick. It is a way of life.

—WILLIAM BURROUGHS, *Junk*

Contents

Acknowledgments

Materials recycled in this book come from various presentations that the entity that claims to be its author—and who usually hides behind the mask of an "I"—had the opportunity to give here and there in the past few years. I thus presented parts of the Introduction at the "Experimenting with Intensities" conference at Trent University, Peterborough, Ontario, in May 2004. I thank Constantin Boundas for this invitation and his great welcome.

I presented parts of chapter 1 at the third international conference of the Centre de Recherche sur l'Intermédialité (CRI) at the Canadian Centre for Architecture in Montreal in March 2001; at the first Gatherings in Biosemiotics, University of Copenhagen, in May 2001; at the annual meeting of the Society for the Social Studies of Science (4S), in Cambridge, Massachusetts, in November 2001; and in the Distortion lecture series hosted by Christine Ross at McGill University, Montreal, in September 2002. The resulting paper appeared first in French in *Intermédialités* and as a chapter in *The Tasking of Identity in Contemporary Art: At the Intersection of Aesthetics, Media, Science, and Technology*, edited by Christine Ross, Johanne Lamoureux, and Olivier Asselin (Montreal: McGill/Queens University Press, 2008). I presented other sections of this chapter at the annual meeting of 4S in Vancouver in November 2006. The resulting paper appeared in *Biosemiotics Research Trends*, edited by Marcello Barbieri in 2007. I presented the final sections of this chapter at the "Science and Belief" colloquium organized by Yves Winkin at *Les Entretiens Jacques Cartier* in December 2007. I thank Yves, Christine, Olivier, Johanne, Éric Méchoulan, and all the CRI participants in my panel; Marcello and all the Biosemiotics crowd, especially Yair Neumann, Kalevi Kull, Don Favareau, Yagmur Denizhan, Jesper Hoffmeyer, Marcella Faria, Tommi Vehkavaara, Luis-Emilio Bruni, Peter Harries-Jones, and Dominique Lestel.

I presented parts of chapter 5 at the Association canadienne-française pour l'avancement des sciences (ACFAS) annual meeting at McGill University in May 2006; at the SymbioticA Seminar at the University of Western Australia in Perth in June 2006; at the Critical Posthuman workshop organized by Ollivier Dyens and Valerie Cools at Concordia University in Montreal in March 2008; and at the Maison des Sciences de l'Homme in Paris in June 2008. I thank Dana Diminescu, Sylvie Gangloff, Christophe D'Iribarne, Matthieu Renault, Matthieu Jacomy, Noura Wedell, Paule Perez, and Françoise Massit-Folléa for the best time ever in Paris since I was born there; Oron Catts, Ionat Zurr and Lilith, Jane Coakley and all the Symbiotic crowd, and especially Guy Ben-Ary and Maya Catts for the good times down under; and Charles Perraton and Étienne Gingras-Paquette, Ollivier and Valérie, Neil Badmington, Charis Thompson, Bart Simon, Jill Didur, Theresa Heffernan, and Crispin Sartwell for the good times at home.

I presented parts of chapter 6 at "The Sinues of the Present: Genealogies of Biopolitics" conference organized by Brian Massumi in Montreal in May 2005. The resulting paper, originally published online in *CTheory*, appeared in *Critical Digital Studies*, edited by Arthur and Marilouise Kroker, from the University of Toronto Press in 2008. I presented yet another part of this chapter at Carleton University in the "Contamination" seminar series organized by Sheryl Hamilton in March 2005. The resulting paper appeared in French in *Les Cahiers du GERSE* in fall 2007. I thank Brian, Sheryl, Arthur and Mari-Louise, Charles, and Étienne for their support and interest in my work.

Thanks to Sarah Choukah, David Jaclin, Ghislain Thibault, Philippe Théophanidis, Claudine Cyr, Lucas Pavan-Lopez, Lidia Cicarma, Marie-Pier Boucher, Caroline Habluetzel, Jorge Zeledon, Ignacio Siles, and all the hack writers who helped me become a better teacher and thus, I hope, a better writer.

Thanks to Line Grenier, François Cooren, Kim Sawchuk, Michael Century, Louise Poissant, Tim Lenoir, Nicolas Auray, and all the other colleagues who have helped make academia a more hospitable habitat.

Thanks to Kim Sterelny, Michael Lynch, Mike Featherstone, John Marks, Arthur Evans, W. J. T. Mitchell, Anne Quérien, Yann Moulier-Boutang, and all the countless editorial board members, anonymous reviewers, friends, and colleagues who provided the necessary opposition—sometimes quite gracefully—to remind me that critical thinking is no consensual work.

Thanks to Roy Britten, Colm Kelleher, Matti Pitkänen, Fritz-Albert Popp, Jean-Claude Perez, and Ruppert Sheldrake, who talked to me, and thanks to Jeremy Narby for not talking to me.

Thanks to Paul Levinson for the "Copyright Notice" story, Bruce Sterling and Rudy Rucker for the "Junk DNA" story, Greg Bear for *Darwin's Radio* and *Darwin's Children*, and Robert Zindell for *Neverness*.

Thanks to Sigur Ros and Philip Glass for the inspiration during the writing hours.

Thanks to Woody Allen for his particular kind of grin from the grim.

Thanks to Tremeur Couix for the word *junkware*.

Thanks to Charles Pless for some linguistic advice.

Thanks to Sylvère Lotringer for a near miss.

Thanks to Jean Baudrillard for the confusion between virality and virtuality . . .

Thanks to the anonymous reviewer for his or her enthusiastic report.

Thanks to Christopher Kelty for a radical review and a follow-up (and more).

Thanks to Cary Wolfe and Doug Armato for their patience and understanding.

Thanks to Fabienne for everything, and the music.

And thanks God for rock 'n' roll.

Coda Lambdas All Over the Place

λ phage is the paradigmatic temperate bacteriophage. A *bacteriophage* is a virus infecting bacteria (E. coli in the case of the λ phage, see glossary entries *bacteriophage* and *phage*). This name was coined in 1917 by Félix d'Hérelle, a French Canadian microbiologist working at the Pasteur Institute in Paris. "Phage" comes from the Greek *phagein* meaning "to eat." λ phage in fact does not eat its host bacteria, in the sense that it does not ingest, digest, and then incorporate it. Instead, it does rather the opposite: the phage penetrates the unicellular organism of the bacterium, replicates inside of it, and later, when its business is done, "explodes" its host. This later phase is called *lysis* (also from the Greek *lyein*, meaning "to separate"), the death of the bacterium by breaking of its cellular membrane. But lysis occurs in only one of the two alternate life cycles of the λ phage, the aptly named *lytic* cycle. λ phage is a "temperate" phage because it can also enter into its *lysogenic* cycle, where its viral genome integrates the bacterium DNA, replicates with it, and quite often remains dormant until conditions deteriorate (and only then will the reproductive cycle kick in, leading to lysis).

Lysogeny was controversial from the start. Félix d'Hérelle did not believe it could exist, and neither did many later microbiologists, including some of the most famous, such as Max Delbrück. Lysogeny could indeed appear counter-intuitive, because it leads to the apparent paradox of a "nonvirulent virus": virulence refers exclusively to the lytic cycle (since only in this cycle is the host cell degraded). "Temperance" then means "nonvirulence," and it took a while for the community of biologists to accept that such a thing could exist in a virus. It gradually happened after World War II, with the works of some of the key actors of the present book, the so-called French connection: André Lwoff and Jacques Monod, at the Pasteur Institute again; see chapter 2, A: "(May) a thousand loops (bloom)."

The λ phage genomic map provides an alternate representation of the structure of the present book. It is, however, no "mere metaphor," nor merely a structural (i.e., formal) point. It is indeed a structural point at first sight: the λ phage genomic map is circular and not unidirectional. Its basic form then is that of the *loop*, which is the main concept of cybernetics, and thus of both computing and molecular biology, and thus of bioinformatics, and thus of the present state of our culture, that is, cyberculture, or, more precisely, "hyperviral culture." The loop is also an archetype of the eternal return, and this is no mere "structural" point, but rather an ontogenetic point, and thus both a physical and a metaphysical point. The loop is the organizing trope and the key bridge between the micro and the macro, the material and the ideal levels of human experience, or, in other words, more in tune with the conceptual framework of the present book, *the molecular and the molar.*

Finding its ontogenetic references in the body of knowledge that today claims authority on the most basic characterization of life itself, molecular biology, the present book investigates its accursed share, the very existence of allegedly insignificant, albeit massive, details on the source of all meaning—DNA itself—inverts it, and returns with a vengeance to "culture itself" as "junk culture." The loop is more than a circle; it is the possibility of more circles inside of the same circle; it is the mother of reflexivity and recursivity. Inside the loop resides the possibility, or maybe even the potentiality, of yet another loop, of a thousand loops. The virus genome does not know a privileged direction, a hegemonic sense: here cohabit sense and antisense; it is all a question of circumstances, of chances and choices in relation to a given, albeit ever-changing, state of the world. In other words, and I insist, it is no mere structural metaphor; the virus (and more precisely the λ phage) is the entity of choice, not only for molecular biology that it helped build, *but for today's ontology and ethics, and hence for current metaphysics.* Today, we late-modern human beings live in tension between our lytic and our lysogenic pathways.

Junkology, if there is ever to be such a "discipline of study," is no heterology, certainly not in the sense given to this expression by a however crucial influence on the conceptual framework of this book, the late Georges Bataille. Junk is not "merely" heterogeneous, as Christopher Kelty helped me realize with his final review of my manuscript. For Bataille, the heterogeneous refers to the sacred, "the highly polarized," and most of his heterodoxy lay in trying to recover some sense of the sacred in some sort of profanation. In other words, heterology was more precisely about TRASH (rather than WASTE), the "merely"—yes, "merely"—heterogeneous, and junkology is ANTITRASH.

Junkology is also about the sacred, but in the sense of Giorgio Agamben rather than in the sense of Bataille. Like Agamben, it wonders about what is left to profane in a world where everything, including signs, is sold, bought, and consumed. The main thesis of this book is that there might be some redemption in junk, if one properly understands what junk means. Yes, we living human beings are now officially junk, but there is some positivity to be found in this rather bleak matter of fact. More precisely, junk might after all be the perfect "object" to help establish an ethics that would be firmly Nietzschean, beyond good and evil, resolutely participating in the transvaluation of all values. "Lambda" is, after all, the most common name for today's form of the singularity, the common singularity that Agamben calls "whatever singularity," and that I consider as today's mode of expression for us disaffected subjects, instances of Homo nexus.[1]

This transvaluation is better understood here as a *transduction,* in both specialized senses of this term: in the metaphysical and epistemological sense given to it by Gilbert Simondon, and in the more restricted sense of virology. Bateson used to say that "we are our own epistemology," and this book attempts to prove him right, in developing a viral epistemology. This is why and how transduction is key.

In the sense of Gilbert Simondon, transduction is "a physical, biological, mental, or social operation by means of which an activity propagates itself from one location to another within a given domain." Simondon's genial insight is to understand individuation (be it physical, biological, or psychological and social) as both action and structure, correlated in the individuating process of propagation by proximity: each structured region of a given domain serves the following region as a principle for its own structuring, in a kind of bootstrapping process (the quintessential loop), "so that a modification extends itself progressively at the same time as this structuring operation."[2]

The sense of the term "transduction" in virology can thus appear as a restricted sense of the Simondonian sense. Here it means the process by which genomic material is transferred from one bacterium to another by a virus, or whereby foreign DNA is introduced into another cell via a viral vector (Wikipedia, "Transduction (genetics)"). The lytic and the lysogenic pathways are thus two alternate modalities of transduction in temperate viruses such as the λ phage. Transduction, in this sense, is a key process harnessed in genetic manipulations; hence it is both a structural analogy for the present book and an aspect of the problem it deals with. In other words, this book operates on a transductive logic, in both matters of structure and contents.[3]

The schematic map of the λ phage genome and of its main transcription pathways thus provides alternate (and junkier) reading trajectories into the book.

There *is* no introduction that comes first, but only a set of "promoters"; the first chapter *is* maybe the only true chapter of this book, in fact a "repressor complex" or, more accurately, "a complex repressor" in both senses of this expression; the rest of the book has been organized into chapters for your comfort only, dear reader, but might better been seen as an orderly display of junk, of "memes" that are indeed "mind viruses," and literally so. In this sense, a "coda" had to come first, and "de-coda" last, unless it could be the opposite (in the same way, this preface may actually be a postface).

Be warned then, this book *is* a temperate virus: it will infect your mind, without, hopefully, getting into an ideal lytic pathway.

The codes accompanying each section of the book refer to the significant elements of the genome of the lambda phage (see glossary entry), molecular biology's beast of choice. Here are (some of) their meanings.

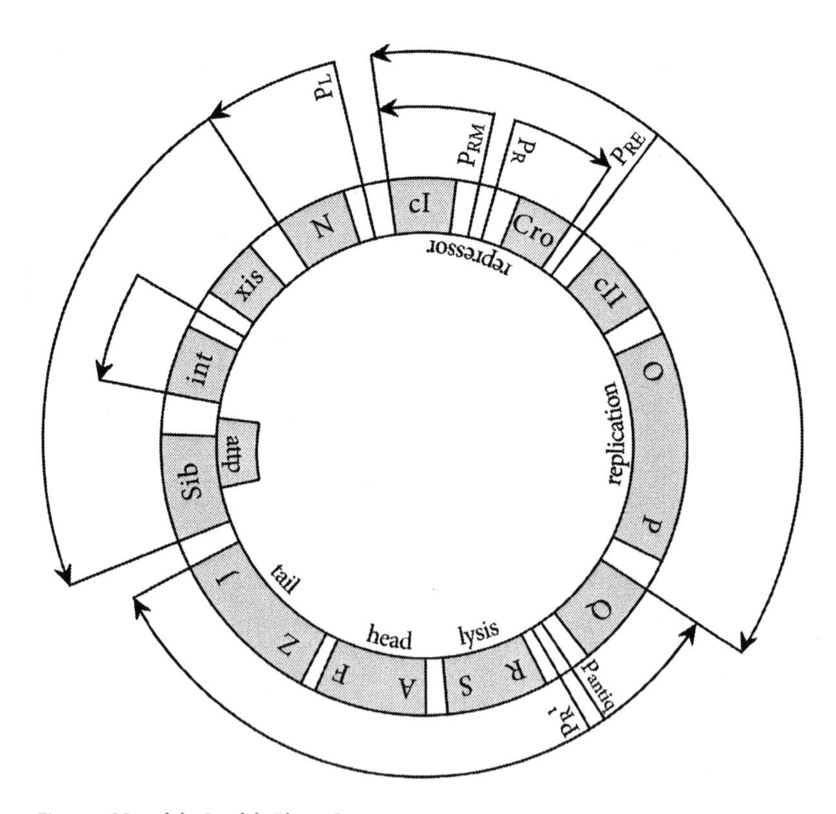

Figure 1. Map of the Lambda Phage Genome.

P_R P_{RE} P_{RM} P_L $P_{R'}$ and P_{antiq} are the promoter sites of the various transcripts associated with the lytic and the lysogenic pathways (the transcripts are represented by arrows on the diagram, clockwise for the former and anticlockwise for the latter; see infra for the explanations of the pathways).

Cro, or Control of Repressor's Operator, is a transcription inhibitor that binds OR3, OR2, and OR1. Cro, CI, and OR1, 2, and 3 form the λ phage repressor complex.

cI or Clear 1 is a transcription inhibitor that also binds OR1, OR2, and OR3.

cII or Clear 2 is a transcription activator.

N or aNtiterminator is a RNA binding protein and RNA polymerase cofactor.

Q is DNA binding protein and RNApol cofactor.

Xis or excision is an excisionase and Int protein regulator, manages excision and insertion of phage genome into the host's genome.

Int or INTegration: the protein product of this gene manages insertion of phage genome into the host's genome.

A, B, C, D, E, F are the genes coding for the head of the virus proteins.

Z, U, V, G, T, H, M, L, K, I, J are the genes coding for the tail of the virus proteins.

S, R are the genes causing the host cell to undergo lysis at high enough concentrations.

O, P are the genes coding for the DNA replication functions of the phage genome.

Sib and attP are not protein-coding genes, but vital conserved DNA sequences.

Introduction, or a Set of Promoters

Robbe-Grillet Cleansing Every Object in Sight, and Vik Muniz Piling Them Up

> We take it that when this state of things shall have arrived which we have been attempting to describe, man will have become to the machines what the horse and the dog are to man. He will continue to exist, nay even to improve, and will be probably better off in his state of domestication under the beneficent rule of the machines than he is in his present wild state.
>
> —SAMUEL BUTLER, "Darwin among the Machines"

P_L *Envoi: When I started this project, I was developing a case of tinnitus (*acouphène*). I needed a challenge for my dazed brain to tame this constant noise inside my ears. I decided to focus on junk. I turned my mind backward to the origins, looking for the ultimate junk, the principle of life itself, a contemporary image of genesis. It led to DNA and its own noise, its noncoding part, dubbed "junk DNA" by molecular biologists.*

This book is thus about junk and about us, late-modern humans. Junk; not trash, not garbage, not waste: junk; mongo, kipple, *gomi*, and all kinds of stuff that grows in stacks and patiently waits for a renewed use. This book is also about now; this particular junction in the anthropogenetic process, this singular moment in time that some claim to be the last moment before the singularity, this unmistakable now just before we might become something altogether different: a whole new species, maybe. This book claims that, in the meantime, we, late-modern humans, are becoming junk, *that is, junkware.* Junk, now, junkware.

Robbe-Grillet Cleansing Every Object in Sight is the title of a painting by the American painter Mark Tansey. He always paints rather arcane, philosophical, or historical scenes in the style of the popular scientific monochrome iconography of the 1950s. Indeed, Robbe-Grillet, one of the founders of *Le nouveau roman*, is represented here in the classic position of the paleontologist cleaning and combing the African desert in search of traces of the origins of mankind, fossils of our ancestors buried in the eroded sands of some formerly green oases of life. Several of Tansey's paintings impressed me a great deal when I saw his exhibition at the Museum of Fine Arts some years ago in Montreal, but none more than the one portraying the French writer, kneeling under a yellow light, brush in hand, frantically cleaning very small objects that litter the sand around him. One of these very small objects is a cow (but this is not the one he is cleaning).

In November 2007 I was reminded of another possible variation on the same idea when I accompanied my son to an exhibition at the Museum of Contemporary Art. It was the Vik Muniz "Reflex" retrospective. Muniz plays with the history of art and re-creates past masterpieces in heteroclite media—Mona Lisa in peanut butter or jelly, Sisyphus in junk—and then takes picture of his copies. His photographs of thread, chocolate, pigment, or color have established him as a leading contemporary artist, or, in his own words, a maker of "the worst possible illusion." His art is a recycling of the visual tradition, or maybe even a re-recycling of it, since he copies and then photographs, very much like Tansey, who photocopies and then paints. The *Mound* series of Muniz's work, displayed on the museum wall as a set of four pictures in the *Reflex* exhibition, really got my attention.

I later learned on the World Wide Web that the four photographs compose a set of 12 × 15 inches laminated place mats made especially for the Printed Matter Company by Vik Muniz. The contents of each mound are printed in pale gray beneath the image, and their list actualizes Robbe-Grillet's small world: Rat Poison, Jade Buddhas, Pectin Drops Meteorites, Taylor Pins, Aspirin, Granola, Silver Glitter, Killer Bees, Dinosaur Dung, Shredded Ezra Pound Canto (XIX), Dice, AZT, Plastic Babies, Cat Hair, Beetles, Brillo Pads, Gold Scrap, Viagra, Jelly Beans, Pubic Hair, Curry Powder, Metal Screws, Edible Worms, California Rolls, Headless Army Men, Coke Caps, Marijuana, Poison Scorpions, Feathers, Wristwatch Parts, Desyrel, Ô de Lancôme. The table is set, and the menu alludes to the strange list of extraordinary animals Michel Foucault once borrowed from Jorge Luis Borges, or to the description of Slothrop's desk at the beginning of *Gravity's Rainbow:* in Thomas Pynchon's words, "a godawful mess,"[1] a junkload.

P_R **Presence of Junk**

Junk is junk: useless or waiting under the falling dust for an improbable novel use for it to be put to. Junk is one step before garbage, its quasi-necessary fate. Junk is garbage ready to happen, trash in the making. On the pavement, junk meets its fate: garbage or garage sale. Junk comes in stacks, piles, drawers, and shelves. Junk lies in the marginal living spaces of the house, garages, attics, and cellars. You forget about it, and it somehow grows anarchically. Junk rusts, fades, decays. Junk lives out the strange life of matter without even a glimpse of a cause anymore. Junk is its own cause.

Junk used to be something else, though. It used to be useful, to serve a purpose, or it was meant to eventually serve a purpose. Its time is always in between, a bubble in the efficient, productive time we unfortunately enough got hooked on in the so-called developed world. Junk lives in a time stasis. Junk is a luxury for the well fed; it incarnates the sentimental scrap we choose to love tenderly in these parts of the world. It materializes the memories of consumption that we grew up idolizing. Junk, on the other hand, is a necessity for the starving, only source of hope and cause of further trouble, chance of recycling and presence of an everlasting lag.

Junk is the order of the day, it now affects communication, invades your (e)mailboxes, floods you with its unbearable way of just being here. You certainly eat it, crave it at times. You must not feed on it, doctors say. Its yard is the flip side of the American dream.[2] Junk is the terrible becoming of the eternal commodity, deprived of use in itself, an incomprehensible sign for a consumer without affect, object of relentless desire for the addict, this half brother of the consumer, the user. Junk is what has been used with fervor and remains because of this fervor. Junk is the nostalgic hope of a renewed fervor to come.

Junk is personal, in the eye (and hands) of the beholder; somebody's junk is somebody else's treasure. Junk cannot be fixed; you have given up on that project a long time ago. Junk must be collected, stored before it is recycled, combined in whole or parts. Junk has to be taken apart, at best. Brought down to its intimate components, exposed. Junk affords no shame. Junk laughs at its own fate, oblivious of time and space. Junk rules.

The more I thought about it, the more I felt that junk was one of the signatures of this time and age. To relocate it on the Master Molecule, *qua junk DNA*, fed back from the realm of the merchandise to the very fiber of our being, was the surest way to close the loop on our idealist mind-set. Then I thought, what if recycling after recycling, re-mediating after mediating, ended up recycling us? What if buying and selling everything, exchanging everything, assigning value to everything, ended up buying and selling us, exchanging us, assigning value to us? Slaves, all of us, but a new kind of slave: enslaved in our code itself. Then I thought, aren't everywhere unemployed workers treated as junk? Aren't we all potentially junk, disposable and recyclable labor force, disposable and recyclable consumers, disposable and recyclable spectators? What if designed obsolescence, after having established its hegemonic hold on the production process, had no other future than being fed back to the production of consumers themselves? Everybody used to be a user; now everybody needs to be used.

Into what kind of units of value are we measured exactly, when our only worth becomes our code? What if living money was not only this money needed to fulfill our basic needs, to reproduce our labor force, but the other way around: as the chicken is just the means for the egg to make more eggs, what if we were nothing but living money, that is, means for money to make more money? I have known this joke since my teen age: "Earning a living? I don't need to earn a living: I am already living." Well, I am not sure anymore ... because I also remember Rachel's famous words in *Blade Runner:* "I'm not in the business ... I am the business." What if we were becoming, in fact, the sexual organs of money, the genitalia of capital?

I am twice deeply indebted to Samuel Butler for the previous paragraph. It was he who first intuited the inversion between chicken and egg, even if he had the modesty to write "it has, I believe, been often remarked that a hen is only an egg's way of making another egg."[3] The idea that humans could indeed been seen as the sexual apparatus of machines comes from "Darwin among the Machines," a letter he sent on June 13, 1863, to the editor of the *Press*, Christchurch, New Zealand, and which eventually became included in his great novel *Erewhon.* Then, long before the age of biotech, but already well into the age of breeding, Butler wrote:

> What sort of creature man's next successor in the supremacy of the world is likely to be. We have often heard this debated; but it appears to us that we are ourselves creating our own successors; we are daily adding to the beauty and the delicacy of their physical organization; we are daily giving them greater power and supplying, by all sorts of ingenious contrivances, that self-regulating, self-acting power which will be to them what intellect has been to the human race. In the course of ages we shall find ourselves the inferior race. ... The fact is that our interests are inseparable from theirs and theirs from ours. Each race is dependent upon the other for innumerable benefits, and, until the reproductive organs of the machines have been developed in a manner which we are hardly yet able to conceive, they are entirely dependent upon man for even the continuance of their species. It is true that these organs may be ultimately developed, inasmuch as man's interest lies in that direction; there is nothing which our infatuated race would desire more than to see the fertile union between two steam engines.[4]

That machines might eventually breed under the form of difference engines rather than in that of steam engines does not diminish the power of Butler's anticipation. Moreover, everything I borrowed from Butler is summarized in his great notions of the potential, and the actualization of the potential as "change." He already knew that "joining and disjoining are the essence of change,"[5] but also that "words are like money; there is nothing so useless, unless when in actual use . . . the coins are potential money as the words are potential language, it is the power and will to apply the counters that make them vibrate with life; when the power and the will are in abeyance the counters lie dead as a log."[6] To this our postindustrial age will have eventually only added the immaterial "information," as it will have de-phased the steam engine into the difference engine; but the "power," the "will," and the "vibrating life," the "joining" and the "disjoining" might appear to be the same, eternally returning under new forms. This, I felt, was also Michel Serres's conclusion when he wrote: "Money is indeterminate, it is everything, a kind of general equivalent, it is nothing, a kind of blank meaning. Information, as blank meaning, is in the process of taking its place, as general equivalent."[7] And thus my intuition took shape. Code is today's general equivalent, I thought, not information. Code itself, one step higher than information on the scale of abstractions. Hereditary information, as the invention of genetic capital, now grounds absolutely the general equivalence in the realm of the living, making us living money in essence:

> We are the eternal return of the enslaved stock under its new name, *junkware.*
> We are the medium, the message, and the code.
> We are worth our code, not worth a rush.

Could calling "junk" a not yet understood, albeit major, part of our DNA eventually amount to making junk out of us? This might have been, I am still afraid, a bad case of syllogism: if DNA makes man, and DNA is (nearly) nothing but junk, could man be, essentially, junk? This could only be hallucinatory delirium, or play on words: assign too much importance to the words, to names . . . and yet . . .

P_{antiq} Philology of Junk

In the fourth edition of the *American Heritage Dictionary of the English Language,*[8] the word *junk* is so defined:

> NOUN:
> 1. Discarded material, such as glass, rags, paper, or metal, some of which may be reused in some form.
> 2. Informal a. Articles that are worn-out or fit to be discarded: broken furniture and other junk in the attic. b. Cheap or shoddy material. c. Something meaningless, fatuous, or unbelievable: nothing but junk in the annual report.

3. Slang : Heroin.
4. Hard salt beef for consumption on board a ship.

TRANSITIVE VERB:

Inflected forms: junked, junking, junks. To discard as useless or sell to be reused as parts; scrap.

ADJECTIVE:

1. Cheap, shoddy, or worthless: junk jewelry.
2. Having a superficial appeal or utility, but lacking substance . . .

ETYMOLOGY:

Middle English jonk, an old cable or rope.

This information is completed by the history of the word, which reads as follows:

> The Middle English word jonk, ancestor of junk, originally had a very specific meaning restricted to nautical terminology. First recorded in 1353, the word meant "an old cable or rope." On a sailing ship it made little sense to throw away useful material since considerable time might pass before one could get new supplies. Old cable was used in a variety of ways, for example, to make fenders, that is, material hung over the side of the ship to protect it from scraping other ships or wharves. Junk came to refer to this old cable as well. The big leap in meaning taken by the word seems to have occurred when junk was applied to discarded but useful material in general. This extension may also have taken place in a nautical context, for the earliest, more generalized use of junk is found in the compound junk shop, referring to a store where old materials from ships were sold. Junk has gone on to mean useless waste as well.

Note the nautical metaphor. Junk becomes the common name for "useless waste" through generalization, but the inverse proposition holds that the useless—albeit possibly recyclable—was equated to "old cable on a ship" through metonymy. The etymology and history of the word is more precise in the online Etymology Dictionary:[9] "'worthless stuff,' 1338, junke 'old cable or rope' (nautical), of uncertain origin, perhaps from O.Fr. junc 'rush,' from L. juncus 'rush, reed.' Nautical use extended to 'old refuse from boats and ships' (1842), then to 'old or discarded articles of any kind' (1884). Junkie 'drug addict' is attested from 1923, but junk for 'narcotic' is said to be older. Junk food is from 1973; junk art is from 1966; junk mail first attested 1954." This points to a Latin root, *juncus*, translated into English as "rush":

> [1]rush
>
> Etymology: Middle English, from Old English rysc; akin to Middle High German rusch rush, Lithuanian regzti to knit

Date: before 12th century: any of various monocotyledonous often tufted marsh plants (as of the genera Juncus and Scirpus of the family Juncaceae, the rush family) with cylindrical often hollow stems which are used in bottoming chairs and plaiting mats.[10]

The First Hypertext Edition of the *Dictionary of Phrase and Fable* by E. Cobham Brewer[11] (from the new and enlarged edition of 1894) specifies the Latin root of the word: "juncus, from *jungo,* to join: used for binding, making baskets, mats."

Juncus is the proper name of junk, whose common name is rush. Its meaning makes it the prime agent of conjunction, of joining together. It is therefore quite perfect that its metonymical referent be a cable. Further philological inquiry also indicates that the Latin *jungere* is only one root for conjunction, and that its etymology goes back even further to the proto-Indo-European root/stem: *yug-, meaning to bind, to harness.[12] The same reference adds that "junk [is] another term from the cattle-breeding lexicon of ancient Indo-Europeans. This word was used only for harnessing cattle into the yoke, so the very word *yoke* is a clear derivative. We can even say for sure that it was neuter in gender in Indo-European: *yugom."

Here is the philological set of transformations in the derivative meanings of the word *junk* from a neuter proto-Indo-European root/stem belonging to the lexicon of cattle breeding to a nautical metonymy generalizing the word to any kind of useless waste. The main operator of the transformation is botanical: Old English gets it from the Old French, where it still refers to a monocotyledon genus (grass, rush): *jonc* in modern French. In modern botanical taxonomy,[13] juncus is "a large genus, with species that hybridize readily and are often difficult to tell apart." Juncacae is a worldwide family of some three hundred species. They are grass-like annuals or perennials, often growing from a creeping rhizome, mostly found in swampy places. There the useless link proliferates in conjunctions, pure spiral line oscillating between sky and underground, a silent prayer to the proper name it calls for: *Juncus effusus Spiralis.*

Juncus is the quintessential rhizomatous genus. A rhizome is a persistent underground stem providing a means of vegetative propagation. A rhizome is essentially neuter and paradoxical: the difference between the rhizome and the stems is external to them; one is underground, and not the other. Could it be only one stem? Does the plant care? It is time, I guess, to invoke the words of the Prophets of the Rhizome, Gilles Deleuze and Félix Guattari:

A rhizome has no beginning or end; it is always in the middle, between things, interbeing, intermezzo. The tree is filiation, but the rhizome is alliance, uniquely

> alliance. The tree imposes the verb "to be," but the fabric of the rhizome is the conjunction, "and . . . and . . . and . . ." This conjunction carries enough forces to shake and uproot the verb "to be."[14]

A Thousand Plateaus is the substratum of my project. I first read it years ago, and I felt that I hallucinated in front of its incomprehensible beauty. From its initial insignificance, it slowly reconfigured my mind, subconsciously. I only started understanding it, I think, the day I understood what I was looking for in junk. Only then did I understand that Deleuze and Guattari provided a set of topological principles to characterize my rhizome of interest, junk:

> 1 and 2. Principles of connection and heterogeneity: any point of a rhizome can be connected to anything other, and must be. This is very different from the tree or root, which plots a point, fixes an order . . .
> 3. Principle of multiplicity: There are no points or positions in a rhizome, such as those found in a structure, tree, or root. There are only lines . . .
> 4. Principle of asignifying rupture: . . . A rhizome may be broken, shattered at a given spot, but it will start up again on one of its old lines, or on new lines . . .
> 5 and 6. Principle of cartography and decalcomania: The rhizome is altogether different, a MAP AND NOT A TRACING . . . Perhaps one of the most important characteristics of the rhizome is that it always has multiple entryways . . . A map has multiple entryways, as opposed to the tracing, which always comes back "to the same." The map has to do with performance, whereas the tracing always involves an alleged "competence."[15]

I thus made these principles mine to draw a "double articulation," a square map: Juncus is the proper name of this rhizome, this map that I intend to map. For a starter, it would be pertinent, I felt, to follow Deleuze and Guattari and look at junk at the *molecular* level.[16] Could it be with this kind of connotation in mind that we could understand this noncoding part of DNA, thus aptly named junk DNA?

A lot of people nowadays are under the impression that we have decrypted the genome, and this is almost true; but most of them also seem to believe that we therefore understand DNA, and that is plain wrong. Most people think that since we have given a structure to DNA[17] and a Central Dogma[18] to the discipline of molecular biology, we can eventually read the Book of Life, and could therefore start rewriting it: the Book of Life is becoming a palimpsest, right before our eyes.

Crick's Central Dogma postulates that heredity works on a one-way transfer of information from DNA in the nucleus to proteins in the cytoplasm of each living cell. From DNA memory to body made of proteins, most people would agree. This DNA memory is thought to be like a Turing machine, a universal computer: an infinite ribbon of boxes, here forever, waiting to be checked and unchecked. Evolution by chance, single mutation—read/write error, accident, bug—is supposed to rule, relayed by necessity, natural selection through adaptation, competition and survival of the fittest. Nobel prize laureates (Monod) and cyberneticists (Bateson) agree on this beautiful stochastic system.

Most people do not even realize that the so-called genetic code actually encodes protein synthesis *on only 1.5 percent of the human DNA bases.* So, there is this beautiful natural process that functions like a medium—following Shannon's model—that encodes and decodes the signals of life itself, hereditary information . . . and it carries about 98.5 percent noise. This noise has been dubbed "junk DNA." Not many people actually know who is responsible for this name.

This metaphor is still the dominant way to portray DNA. Thus DNA is often compared to a book, such as on the Human Genome Project Web site, where it is compared to "200 Manhattan telephone books" (of one thousand pages each). We are told that it would take twenty-six years to read it . . . In this metaphor's framework, junk DNA is seen as the set of numbers that are no longer in service (and this for all the possible numbers that have ever been since time immemorial!).[19] Junk is fossil or selfish genes, some say; silent, all agree. By that they mean not transcribed nor expressed in proteins; inoperative; out of order; garbage, an awful lot of it. I started having doubts about this picture. I felt that "junk" was more than just a convenient name for what is yet unnamed. If it were merely useless, after all, why not call it "trash" or "garbage"?

I decided to observe and map on the World Wide Web the semantics of the discursive formation of Junk DNA. If one were to make a junk map today, I figured, he or she would best start with the Lightest Pile of Junk of this Earth, the World Wide Web.

P$_{RE}$ **Mapping Junk**

I thus started the project on the World Wide Web. I used Google™, the most efficient detector of immaterial scrap in the world. I entered the words "junk DNA" in its little window, and then many other words. With a crew of five graduate assistants, I surfed, chartered, and explored the shores of distant rings of virtual space junk for three years. We drew maps. We gathered an archive, started blogs and yahoo groups and whatnots. We carried it out in an iterative process, mixing formal and informal procedures. We collected over five hundred bookmarks about the work of a few dozen individuals working in the sciences (chemistry, biophysics, bioinformatics, anthropology, medicine), engineering, and science fiction. The criteria for significance were ad hoc characterizations of the originality, fecundity, and metaphorical power of the works displayed on the Web. We paid no attention whatsoever to the truth-value of the claims found on these Web pages.

We started with a scheme of the lexical organization of our archive derived from Greimas's semiotic square. It seemed obvious to start from such a methodological basis since Greimas claimed that the semiotic square, whose concept is based on the notions of conjunction and disjunction, is the elementary structure of meaning. In fact, Greimas's insight is crucial here because he distinguished *two* modes of disjunction: contrariety (or opposition) and contradiction. Both suppose that two terms are mutually exclusive, but contradiction requires exhaustivity, when contrariety does not.[20] A third type of relation between concepts or "semes" (S) is *complementarity* or implication, two alternate names for conjunction. The three types of relations are given in the basic scheme in Figure 2.

An informal browsing of our first archive led us to three basic semiotic squares, around the following couples of "semes": silence/language, fossil/program, structure/medium. These three forms of contrariety seemed at first to exhaust the possible lexical oppositions relating to junk DNA. The first and the most elementary square questioned the basic assumption that junk DNA, since it did not code for protein synthesis, was actually "silent," that is, meaningless or functionless. In contrariety to this notion, it develops on the notion that whole DNA could be structured as a language. The second square referred to the hypothesis according to which junk DNA was mostly composed of repetitive sequences and pseudo-genes, that is, once-functional genes discarded through the evolutionary process. In contrariety to this notion, it develops on the notion that whole DNA could be understood as a programmable machine (i.e., a Turing machine). The third square, finally, questions the notion that noncoding or junk sequences play a structural role, if any. In contrariety to this notion, it develops on the notion that whole DNA could be understood as a medium, of which heredity could merely be one modality. The overall scheme produced by the composition of these three elementary squares is shown in Figure 3.

When we focused on the positive characterizations, it soon became quite obvious to us that the discursive formation of "junk DNA" could be adequately described as the confluence of four basic vocabularies: first- and second-order cybernetics, evolutionary theory, and what, for lack of a better term, we called "metaphysics." The first two vocabularies appeared true to the metaphorical origins of molecular biology. For the first vocabulary, it would mean using the

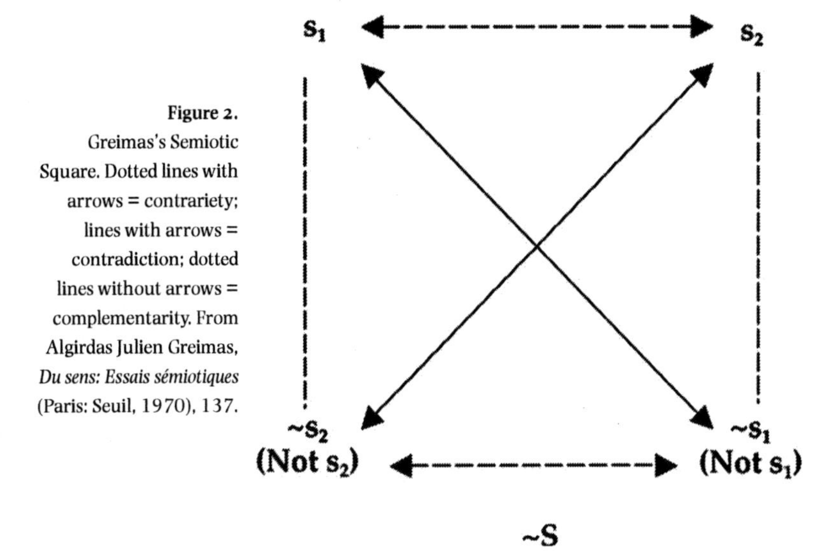

Figure 2. Greimas's Semiotic Square. Dotted lines with arrows = contrariety; lines with arrows = contradiction; dotted lines without arrows = complementarity. From Algirdas Julien Greimas, *Du sens: Essais sémiotiques* (Paris: Seuil, 1970), 137.

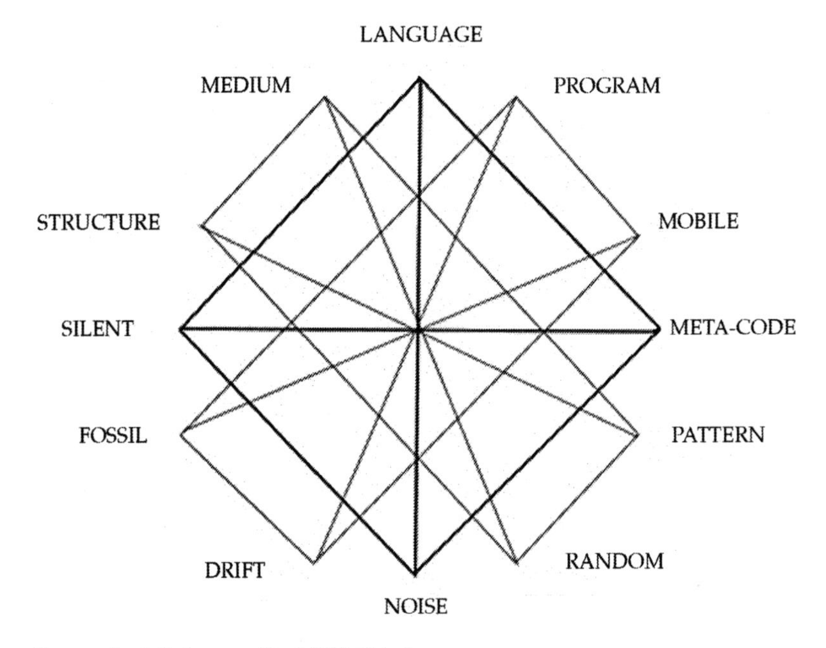

Figure 3. Semiotic Squares of Junk DNA, Take 1.

category of "noise" to refer to the noncoding sequences, and for the second-order cybernetics vocabulary, it would mean looking for emerging patterns or "order" out of seemingly chaotic or complex sequences. The third vocabulary meant looking at the noncoding sequences from the standpoint of evolution theory, that is, from inside the standard neo-Darwinian paradigm. The fourth vocabulary, finally, meant focusing on questions at odds with the three previous vocabularies, mostly in a teleological manner: the question about a specific function or sets of functions for noncoding sequences was key to this fourth lexical quadrant. Eventually, we proposed the model in Figure 4, inherited from our previous three-squared semiotics, for the core of our semantic network.

The southern and northern triangles, respectively, referred to first- (information message code) and second-order (complex program pattern) cybernetics vocabularies, the eastern triangle to the vocabulary of evolutionary theory (evolution random mutation), and the western triangle to the vocabulary of metaphysics (function language design). Each of these triangles was in our mind the organizing seme of a specific vocabulary denoting a specific point of view on what "junk DNA" could do or be. Through iterative and informal

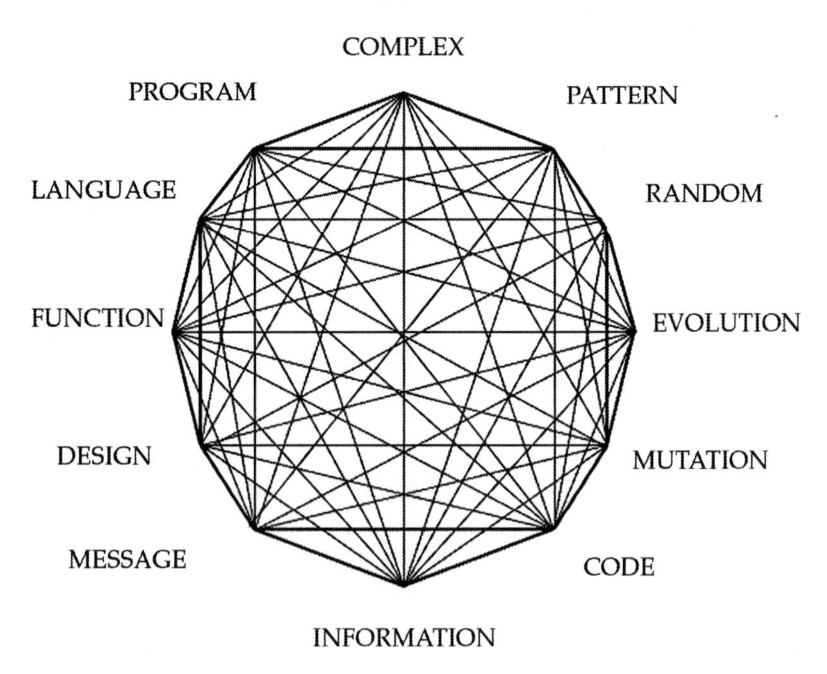

Figure 4. Semiotics of Junk DNA, Take 2.

queries on our online archive, we eventually complemented these basic concepts to create a full-fledged lexicon including both negative and positive as well as specialized characterizations of "junk DNA." After much exploration and analysis, we opted for a final version of the lexicon including forty-nine terms: *information, function, computer, structure, light, complex, evolution, software, code, language, chemical, design, memory, program, quantum, mutation, random, message, coding, intelligence, virus, pattern, origin, purpose, god, plan, regulation, conscience, sign, mobile, introns, noncoding, repetition, repair, fire, eukaryotes, fossil, parasite, useless, jump, medium, silent, noise, selfish, cosmic, drift, electric, transposons, automata.*

We then performed online queries for the co-occurrences of "junk DNA" and the terms of our lexicon for six consecutive months in 2004, once in 2005, and once again in 2006. Using various softwares for network analysis and visualization, we obtained a set of maps allowing us to portray in a dynamic fashion the various relationships between the founding vocabularies. Although these maps were very helpful tools at first to describe the archive, they eventually became a meaningless exercise in statistical mania. As such, they provided a general and preliminary picture of the subject. But we also used the maps and the semantic analysis to decide which interviews to carry out: they allowed us to pick our subjects in order to cover as much semantic real estate as possible. That indeed we did, from the most orthodox of the heretics to the plainly ludicrous (or so it could appear). We extracted a preliminary list of twenty individuals contributing significantly to our archive. I interviewed the first six of them.

In fact, I interviewed five, or maybe even six, alleged *cranks.* According to Wikipedia, this term is "a pejorative term for a person who writes or speaks in an authoritative fashion about a particular subject, often in science, but is alleged to have false or even ludicrous beliefs." I was unconsciously drawn to interview people condemned as such from the other side, the good side of scientific reason and precise writing of matters of fact. They were, however, the best persons from whom to expect some extraordinary insights on junk. They allowed me to chart the risky waters of the blind side of the cybernetic paradigm.

I had decided to locate this project, the site of its intervention, at the interface of science and fiction. The methodology of my project followed—albeit with a twist—the principles of the strong program in sociology of knowledge: causality, impartiality, symmetry, and reflexivity.[21] I was thus causally, impartially, symmetrically, and reflexively drawn to marginal contributors to mainstream science: scientific outcasts, not fiction writers, but experts in junk, *junkers,* if you will. I figured that if 98.5 percent of the bases of our DNA have been declared to be useless, selfish junk by the highest scientific authorities, only marginal contributors would dare to claim expertise upon its knowledge. Here was my own take on (reverse or circular) causality. I decided not to care at all for the relative truth or rationality of the theses that I would encounter in my exploration of current theories and speculations about junk DNA. Here was my own take on impartiality (and symmetry, which pretty much amounts to the same principle, in my mind). In the world of fictions, I was looking for performativity, not truth or reason. This methodological flattening is crucial to my research endeavor. This, I insist, was only a methodological principle, and in no way an ontological axiom.

I could (and maybe should) have gone on with the list of interviews. Instead, I decided to start writing. I was driven by a sense of emergency. Each day, on a sign in the street *(Biotica; Mystical Products),* or on TV, I could find more evidence of the presence of junk. I suddenly awoke to a culture of junk. And junk, I realized, was irreversibly transforming us into *Homo nexi,* the disaffected subjects of its culture. I had focused on *junk DNA*—as if the compound expression was but one word—and soon realized that junk was a much broader concept. I collected a lot of scrap in the process and I progressively accepted that it was irreversibly transforming me into a scrap scholar (so much for reflexivity). At a time when I, like my iPod, oscillated between shuffle and repeat, it produced this book, which was first another book.

$P_{R'}$ **Biomolecular junk**

First, I examine the dominant scientific discourse that attempts to define us, now: molecular biology. I find and lose sight of junk on our most intimate fiber, DNA. I follow the ups and downs of this part of DNA, the huge majority of its bases, in fact, that seemed at first to resist the standard paradigm. DNA stores our genetic memory, does it not? It is the medium of our heredity; it is organized according to the codes of codes, the code of life itself, isn't it? Then how to explain that in the human, as in all multicellular forms of life, most of the DNA bases do *not* code for proteins? How to explain this waste in the natural economy? If DNA is indeed such a living medium, how to explain, then, that its signal-to-noise ratio is under 2 percent?

Molecular biology, like computer science, and perhaps even more than computer science, is the ultimate becoming of this twentieth-century synthesis of thought called cybernetics. This first part of the book tracks back the evolution of the cybernetic understanding of life and examines it from the standpoint of its blind spot, that is, *junk DNA.* It builds on recent history of molecular biology, on the works of such great scholars as Lily Kay, Evelyn Fox Keller, Judith Roof, or Richard Doyle that it updates and questions. Although all of them have developed enlightening theses on the cybernetics metaphor applied to life itself, none seem to have bothered so far with what seems to amount to the breaking point of all these founding analogies: junk DNA. Building on these narratives, *Junkware* attempts to rewrite history from the standpoint of what was first left out, this uncanny "detail," the presence of junk on DNA. What, therefore, does life look like when we look at molecular biology from its blind spot?

Three chapters compose this first part of the book. The first chapter, "How Junk Became, and Why It Might Remain, Selfish," anchors my project in the (bio)semiotics of junk, understood as an inquiry into the cybernetic metaphor applied to the understanding of life, its modes of reference, and the question of "genetic insignificance." Parting ways with the standard historical narratives of molecular biology, I chart my journey between two opposite and seemingly incommensurable takes on junk DNA, that of Richard Dawkins and Jeremy Narby. Then I proceed by coming back to an argument in the pages of the journal *Nature*, in 1980, to show how and why Richard Dawkins's notion of "the selfish gene" was then wrongly extended to found a theory of selfish DNA. The chapter relates this issue to the question of teleology and the historical debate over the inference of design, and establishes the epistemological and biopolitical contours of the subject. It thus establishes that the standard model of molecular biology, centered on Crick's Central Dogma and its ubiquitous cybernetic metaphor, extended by Dawkins's selfish hypothesis, is but one way to characterize junk DNA, and a troubling one at that, logically flawed and established in an authoritarian way. The remaining chapters of this first part then proceed to describe an alternative account of biomolecular junk, pre- *and* postdating Crick's decision.

Chapter 2 thus describes the moves from "garbage" to "junk" DNA in the 1960s and 1970s, and what junk slowly became when it entered the realm of simulation, in the late 1980s and early 1990s. In other words, I consider the junk becoming of DNA on both sides of the dreaded 1980s, object of the first chapter. I go back to the question of teleology and demonstrate how its cybernetic redefinition as *teleonomy* allowed for the emergence of a renewed discourse on genetic regulation, where junk DNA could have a singular importance. I focus in this chapter on the works of two researchers: Susumu Ohno, who coined the expression "junk DNA," and Roy Britten, who pioneered the studies of potential regulatory functions buried in junk. I then move to the late 1980s and early 1990s and explain what happened to junk DNA once the field of bioinformatics was created and DNA entered the hyperreal and life became understood as a software problem. I argue that the progressive availability of hardware such as automatic sequencers, techniques and software such as recombinant DNA, and computerized handling of sequences progressively made wetware appear obsolete. From then on, molecular biology was more concerned with databases, bits, and models; a deluge of data fell on the synthetic discipline, and made it the perfect convergent industry for biocapital to thrive on. In this process, junkware became the quintessential gold mine of new breed of capitalism: it came to be the *aufhebung* of code itself.

Chapter 3 provides contemporary alternative—and possibly even pseudoscientific—accounts of junk DNA. It nevertheless offers different metaphors to frame our understanding of the noncoding parts of DNA, away from the cybernetic metaphor of the program, object of the first two chapters, and into those of the field, the antenna, and the network: or more broadly into a semiotic regime where DNA is best seen as a multimedium and life is cast as an interface problem. I return in this chapter to the other bootstrap of my research, and track back the set of references and the cast of characters associated directly or indirectly with the ideas presented in Jeremy Narby's *Cosmic Serpent.* I show how a fuzzy configuration emerged among the marginal thinkers, the kooks and cranks of revolutionary science; a configuration linking fuzzily light emission and reception on DNA (biophotons), internested morphogenetic fields, from the quantum level to populations, epigenetic processes, and, why not, a whole mathematical world of strange entities and paradoxical phenomena. Here junkware engulfs the whole universe, analog and digital, in power and acts, material and ineffable, organic and mineral. Here the analytic engine between my ears comes to a halt, and opens to the realm of experience and feeling. Dazzled and laughing, bursting into a momentary suspension of skepticism, I let the hyperreal rush take over my inspiration and rhythm my prose. I end up in Las Vegas, following the meanders of what already amounts to a culture of junk, guided by a real-world hybrid of Dana Scully and Fox Mulder, into the strange equivalences of abduction and near-death experience, light body and werewolves, anomalies of all kinds. This, my own *X-file,* sets up the table in turn for another archaeology of junk, in its molar dimension this time.

PRM The Junkness of Culture

Once I have established in great detail the meaning of junk to life and, most important, the meaning of life to junk, I turn to an entirely discarded metaphysics of science and life, and revel in this junkness of being by tracking it through culture, philosophy, and the future. Here, I engage into a posthumanist journey, where "the posthuman" is equated at first with a becoming-junk.

Many of our contemporaries seem to believe that the human species is now entering the phase of the production of its own metamorphosis. Our cultural landscape is littered with promises and prophecies, and the stakes are high for whom to speak the louder, for whom to capture best the gloom and doom, or, alternatively, the hopes and dreams, of a humanity left shaking by the twentieth century. Those who today try to make us believe that they will soon be able to synthesize a whole human being from a bunch of chemicals plus information agree in principle with those who ban reproductive cloning and make it "a crime against the species." None seems to doubt the scientific premises of the whole issue. All feel that preventing or aiming at the cloning of an individual considered genetically identical to another human being will not hinder, but rather will facilitate, the cloning of parts, sequences, cells, or organs of human beings. On one side, the ban on reproductive cloning provides the moral grounds that reassure the masses about the seriousness and integrity of those in charge, while, on the other hand, the cultural folklore about human clones reinforces the feeling that we are—or soon will be—able to do it. In both cases, the agenda for the progressive commoditization of human synthetic matter is further advanced and the reign of the living money made nearer.

In the West, we are sold every day the promises of a better health thanks to biotechnological fixes, at the exact same time that we witness the global crumbling of the welfare state, and most of all of its promises for universal health care. In fact, we slowly enter the era of the mass production of undead beings, zombies and goleymes. The mass production of cadavers, to quote Heidegger, an expert in this notion, is slowly but surely being replaced by the mass production of undead beings. By this I mean more than a horror/science-fiction trope of rhetorical power over the imagination, but quite literally the production of living entities from human origins, but with the legal and cultural status of dead matter.

Instead of locating the issue inside some disciplinary ghetto, be it philosophical, *Junkware* argues that there cannot be a discussion of our forthcoming posthumanity without a clear understanding of its scientific basis. It argues against bioethics, against any a priori moral posture that would not understand, discuss, and doubt in details the scientific "matter of facts" that allegedly support it. It argues for such an understanding of the evolution of the standard paradigm of molecular biology and intends to show that its shortcomings were, and, to a certain extent, still are, the very basis of our cultural landscape. In other words, *Junkware* demonstrates that "junk" is both the sign that something went wrong with molecular biology and the master trope of our current (Occidental and hence global) culture.

Junk, this incarnation of the perpetual potentiality of recycling, is today's best figure of the virtual, and our whole culture has indeed become virtual. We might not be cyborgs or posthumans yet, but digital beings indeed we are: our age is this age when a major metaphysical concept has conflated with a new and limitless real estate for capitalism to grow wild, when the virtual and the cyber have merged and become indistinguishable. It is thus no meaningless coincidence that junk was the name given to the allegedly useless part our most intimate fiber, the repository of our genetic inheritance, as it is the name of our most addictive product and of our worst food. It is no meaningless coincidence that somewhere in the process of creating this culture of junk, we, the last humans before the singularity that might regenerate our species, became junkware.

"Junk," then, is no concept; it is, rather, one of the signatures of this age.[22]

So, "junkware" is the name I chose to give this ordeal, turning the modern industrial and postindustrial excretions into a new sense of what being human can mean, now. I am no moralist, even if most of this book has an ethical aim, in the sense the late Foucault gave to ethics—that is, the reflexive practice of freedom[23]—and shares his distaste for morals of all kinds. I do not think that regulation stems from disciplining and punishing. Ethical committees and bans of all sorts might at best slow down, one case at a time, the dehumanizing process of turning life into a commodity, but will in no way end it. I am no proselyte either; I do not advocate the opening of the doors of perception. I do not adhere to any party or personal development movements and I have no libratory agenda. Redemption is a personal matter, and the definition of the person is what is at stake now, as always, but perhaps a bit more than usual.

Junkware is about our contemporary cyberculture; a specific *culture,* as in tissue culture *and* pop culture. I argue, then, that it is essentially a culture of junk (and not of trash, as formerly thought). Strange characters and trivial objects populate the second part, linking humanity and its future, cultish science-fiction authors and lonely terrorists, stylish architects and writers-junkies, avant-garde bioartists and French philosophers with you and me, today's disaffected subjects, the targets of viral marketing, bad eating habits, spam of all kinds, addictions of all types. Junk is the organizing principle of that which cannot be organized, the operating mode of that which has no function (yet). Junk is, and, I claim, junk rules. Welcome to a culture of junk.

Three more chapters compose the second part of the book, titled "Molar

Junk: Hyperviral Culture." Chapter 4, "Close Encounters of the Fourth Kind," discusses the metaphysics and ethics of posthumanity. It starts from an extended discussion and an update of Deleuze and Guattari's *A Thousand Plateaus.* I claim that our societies are moving out of control (as they had previously moved away from discipline), and into what I dubbed the era of the machine of the fourth kind, genetic capitalism. Sequences, genes, cells, and organs are becoming the new commodities, embodying a bright future for the extension of the market. If today's global economy is under the spell of "One Market under God" (Thomas Frank), gene sequences and other living codes will be its junk bonds, objects of the new risky and high-reward markets of genetic capitalism. The French legislative invention of "the crime against the species" and several other contemporary phenomena, all dealing with biotechnological and biopolitical innovations, beg for an extension of the Deleuzo-Guattarian framework, revisiting the central metaphysical notions of "common nature" and individuation. By that I also mean that Deleuze and Guattari's work, and especially their *Anti-Oedipus,* is a part of cyberculture qua junk culture: a culture that sees cyberartists of all kinds quote "The postscript on societies of control" as much as Benjamin's "Work of art in the age of mechanical reproduction" or Duchamp's *Grand verre;* a culture where everybody is taking care of his or her very own body without organs; a culture that elevates the *ritournelle* to the status of most envied commodity, and finds rhizomes everywhere it looks. Junk is the fabric of this proliferating rhizome, and as much the stuff of their *molecular unconscious* as of a molar *appareil de capture.*[24]

Following this line of flight, chapter 5 examines the transitory form that we, late-modern human beings, might embody for the posthuman to come, this progress junky, this self-improvement zealot. I deploy here the philosophical fiction of Homo nexus, today's hyperconnected individual, but also, according to its Latin juridical etymology, the indebted free person reduced to the status of a quasi slave. I thus look at us, remaining link (and soon-to-be *missing* link) before the singularity, as these dis-affected or dis-junked subjects suffering from Promethean angst. That the human is "essentially" junk, however, is not only worrisome or sad. It is also the definite irony of his (mis)measure. For junk is above all a figure of the virtual: that our "nature" could only be virtual now, at this time of pervasive plasticity, should not come as a surprise. This could also mean that it is only now that we can realize that life has always been junk, that is, always potentially recyclable. Maybe the eternal return is best seen as a Woody Allen joke: "More than any time in history mankind faces a crossroads. One path leads to despair and utter hopelessness, the other to total extinction. Let us pray that we have the wisdom to choose correctly." Maybe Klossowski was right, and as the gods died of laughter when one of them claimed to be the only one, humans might die of the same laughter when many of them will claim to (soon) become overmen. Maybe the fact that we are all potentially now in the position to be the ass of the festival is the joke of the day.

Finally, chapter 6, "Presence of Junk," closes this narrative with an alternative genealogy of junk, parallel in time and synchronous in its effects with the genealogy of biomolecular junk established in Part I of the book. At approximately the same time, our junk culture found its full visible appeal in Ridley Scott's *Blade Runner,* the movie, as it had found its true name in Philip K. Dick's fictions of the 1960s. At approximately the same time, Ted Kaczynski, aka the Unabomber, aka the junkyard bomber, started his killing spree; Rem Koolhaas, the disjunked architect, started building our junkspace. Awakened ten years before from the words of William Burroughs, Dr. Junk himself, relayed by those of our first virologist philosopher, Jacques Derrida, the hypervirus infection then reached pandemic proportions. The virus, this ultimate junk code, had burst through the frontier of the digital and gone critical: from then on, there could be no ontology but a viral ontology, and it would eventually prove to be deadly. All these contributions, across the divides between science and fiction, philosophy and pop culture, belong to the same hyperviral culture; they are instances of the junkness of late-modern culture. *Junkware* ends up as the narrative of Homo nexus's autopsy, the declaration of his death by overdose of junk, and the promise of its eternal return.

A final chapter, titled "De-Coda," concludes the book on the false promises of the molecular revolution, the persistence of junk, and the victory of stasis over entropy. It also provides a last word on molecular and molar junk alike, in the form of a *vanishing sequence.*

The fourth wave of critique (after Kant, Marx and Adorno) of the late 1960s and early 1970s convinced me that we entered then into the era of the great inversion. As Guy Debord once stated, "in a *really inversed* world, the true is a moment of the false."[25] Jean Baudrillard concurred and raised the specter of the simulacrum from the (fake) word of the *Ecclesiastes* itself: "The simulacrum is never what hides the truth—it is truth that hides the fact that there is none. The simulacrum is true."[26] In this era, junk can indeed ground a culture, as the ultimate figure of the virtual, a reservoir of always potential ex post facto regulatory framings. In this really inversed world, the beginning is but an end, or, as Hillel Schwartz wrote in the Refrain of his great *Culture of the Copy,* "to write an introduction is to prophesy post hoc."[27] Consider it done.

Part I 3' Biomolecular Junk

Digitality is its metaphysical principle (the God of Leibniz), and DNA its prophet.

—JEAN BAUDRILLARD, *The Orders of Simulacra*

Chapter 1, or a Repressor Complex

How Junk Became, and Why It Might Remain, Selfish

> Men and words reciprocally educate each other; each increase of a man's information involves and is involved by, a corresponding increase of a word's information.
>
> —CHARLES SANDERS PEIRCE, "Some Consequences of Four Incapacities"

On April 14, 2003, the International Human Genome Sequencing Consortium, led in the United States by the National Human Genome Research Institute and the Department of Energy, announced the successful completion of the Human Genome Project, more than two years ahead of schedule. The announcement was greeted with ample press coverage displaying numerous expressions of a wide spectrum of emotions, from puzzlement, wariness, and hype to religious awe. Often compared to the Manhattan Project, or perhaps more cautiously dubbed the Apollo mission of biology, the international effort to sequence the 3 billion DNA bases in the human genome (at a price tag of more than one U.S. dollar per base) is considered by many to be one of the most ambitious scientific undertakings of all time, leading, at last, to the deciphering of the Book of Life. "Today we are learning the language in which God created life" had even anticipated President Bill Clinton at an internationally televised press conference for the announcement of the first draft, two years earlier.

Perhaps the biggest "surprise" of the announcement was the low number of genes revealed by both the public project (31,000 protein-encoding genes) and its private competitor, Celera (26,000). In his commentary on the announcement of the first draft, in February 2001, David Baltimore had already noted that "the number of coding genes in the human sequence compares with 6,000 for a yeast cell, 13,000 for a fly, 18,000 for a worm and 26,000 for a plant."[1] Disappointment indeed, when the number of genes in the human genome was regularly estimated to be more than a hundred thousand before the first draft was completed.

For the philosopher of science, this disappointing surprise recalls Thomas Kuhn's notion of an anomaly, this strange feeling that nature is not playing by the (paradigmatic) book.[2] In this chapter, I argue that the standard paradigm of molecular biology had in fact settled the matter a good fifteen years earlier. I describe an argument that took place in the pages of *Nature* in 1980, and show that a display of authority (and a logical fallacy) allowed relative closure on a certain picture of DNA, a picture of DNA solely devoted to being the carrier of protein-encoding genes. I contend that this argument, in line with a defense of the phenotype paradigm (an alternative name of the standard paradigm), effectively restricted the meaning of the Book of Life on this "prejudiced definition of genes."[3]

The first draft of the human genome[4] had already allowed David Baltimore[5] to characterize the genome as "a sea of reverse-transcribed DNA with a small admixture of genes," with only 1.1 percent to 1.4 percent of its 3.2 gigabases actually encoding proteins, a third transcribed into RNA, and more than half devoted to repeated sequences. Baltimore added: "I find it striking that most of the parasitic DNA came about by reverse transcription from RNA."[6] But the very category that Baltimore used to characterize the majority of DNA bases, "parasitic," stemmed, as we shall see, from the episode studied hereafter. In reopening it for history's sake, I want to reflect on Paul Berg's premonitory questions: "shall we foreclose on the likelihood that the so-called noncoding regions within and surrounding genes contain signals that we have not yet recognized or learned to assay? Are we prepared to dismiss the likelihood of surprises that could emerge from viewing sequence arrangements over megabase rather than kilobase distances?"[7]

It seems highly relevant to me that Berg articulated this interrogation in 1991, ten years after the episode studied in this chapter. To me, it means that Crick and his colleagues actually foreclosed the question for a good ten years (at least, since some practicing biologists would still feel the same way today). For these ten years—the dreaded 1980s—junk actually became selfish. Since 1991, as we shall see later, the field reopened, but, as I will contend, in a relative fashion.

cI **On Genetic Insignificance and Its Semiotics**

In 1977, molecular biologists discovered, much to their surprise, that human genes, like those of all eukaryotes (or higher organisms), can be interrupted (or "split") with noncoding sequences for protein synthesis (also named introns). Two years later, Francis Crick pondered on the importance of this "mini-revolution in molecular genetics," only to conclude that "there can be no denying that the discovery of splicing has given our ideas a good shake."[8] Within the limits of knowledge at that time, this noncoding part of DNA was dubbed "junk DNA." Actually, this name predated the "discovery of split genes," and a researcher named Susumu Ohno had intuited the presence of this "non-sense DNA" and named it since the early 1970s. In fact, he had built this intuition on an apparent paradox. Following the established consensus of the time, he had noted that "there apparently is no physical interruption between one gene and an adjacent one on either side of it," but he also wondered, "how is it then that each gene as a rule transcribes a separate *messenger* RNA?" His conclusion was, by all means, logical: "It must be that the space between adjacent cistrons is occupied by a stretch of *nonsense* base sequence."[9]

For the remaining thirty years or so, however, the ambiguity of this "nonsense," paradoxically located on the molecule harboring the "code of life," remained. In 1997, for instance, a common textbook in the field still described four main "hypotheses" summarizing the numerous attempts to provide evolutionary explanations for the presence of "noncoding DNA." The first hypothesis claims that this "lack of function" is only apparent, that DNA is in fact, "wholly functional," and that so-called noncoding DNA "performs essential function, such as global regulation of gene expression" (Zuckerkandl 1976). The second hypothesis, however, follows Ohno's insight more closely and considers that "the nongenic DNA is useless 'junk' (Ohno 1972), carried passively by the chromosome merely because of its physical linkage to functional genes." The third offers a variation on the same idea and considers this "nongenic DNA" as a "functionless parasite" (Östergren 1945). The fourth hypothesis, finally, also predates the discovery of split genes. It holds that this DNA "has a structural or nucleoskeletal function, i.e., function related to the determination of nuclear volume but unrelated to the task of carrying genetic information (Cavalier-Smith 1978)."[10]

The fourth hypothesis, and the name of one of its main proponents, Thomas Cavalier-Smith, refer in fact to a whole field of study centered on yet another paradox: the so-called *C-value paradox*. The "C-value" of a given genome is another name for its size, that is, the mass of DNA in an unreplicated haploid genome, such as that of a gamete (sperm or ovule).[11] "C" in turn refers to the fact that it is usually constant in any one species. As Cavalier-Smith explains, however, "hardly had this constancy been established when Mirsky and Ris (1951) demonstrated huge interspecific variation in C-values that bore no relationship to differences in organismic complexity or to the likely number of different genes in the species studied . . . As it became increasingly firmly established that genes consist of DNA this puzzling lack of relationship became more and more of a problem, now usually called the C-value paradox (Thomas, 1971)."[12]

Ryan Gregory has noted, however, that this paradox only holds "under the expectation that genome size should be equal or proportional to gene number and should therefore increase with 'organismal complexity.'" He added: "this paradox has literally disappeared with the discovery that genomes contain 'excess' (largely repetitive) DNA that is not transcribed into functional products."[13] Again, Gregory repeats the essential ambiguity of this "excess DNA" but also concludes that if its origin and precise function remain an unsolved problem, it is therefore an "enigma" and not a paradox. This "enigma," however, is of a crucial importance. Somehow, it threatens the very basis of the standard paradigm of molecular biology and perhaps even of evolutionary biology qua neo-Darwinian synthesis. Gregory is right, then, to insist that it reaches back to question the very notion of the genome, a notion initially coined by the German botanist Hans Winkler in 1920 to refer to "the haploid chromosome set," but which soon took *two* simultaneous significations, as the full set of chromosomes *and* all the genes contained therein. As Gregory noticed, the realization of the existence of huge amounts of "excess DNA" made this ambiguity intolerable, and made it clear that "total chromosome content and number of genes are not interchangeable." Moreover, he is right to insist that "this deceptively simple observation underlies one of the longest-running puzzles in evolutionary biology."[14]

At the time of the discovery of split genes, however, the primary reference of the "genetic code" metaphor was so well established—even if "erroneously"[15]—that it wasn't questioned anymore: DNA was to remain the medium for the transmission of an individual's genotype; the way in which the DNA bases are organized *is* a message in a digital code that constitutes the individual's "genetic memory." As we shall see later, a variation of the third hypothesis (on the *parasitic* nature of excess, non-sense or noncoding DNA) allowed, albeit not without some controversy, to build a consensus that eventually salvaged this standard paradigm for a while (the dreaded eighties, again). It did so, however, at the price of bracketing the central question: does this mean that these non-sense sequences, which apparently mean nothing with respect to protein synthesis, mean nothing *at all* in the framework of the code metaphor?

Before I come back to narrating the enfolding of this argument and eventually try to explain how junk actually became selfish, I will now attempt to respond to this question by contemplating the status of the noncode vis-à-vis the code metaphor, in turn considering it in light of the semiotic approaches of Umberto Eco, Roland Barthes, and Jean-Pierre Faye. The choice of these three different semiotic perspectives on the genetic code metaphor comes, in each case, from the coupling of a specific semiotic perspective with the various hypotheses on non-sense DNA presented above.

The standard reply to the question about the possible function of noncoding DNA registers in a neo-Darwinian, evolutionist perspective that integrates quite well with the general economy of the "Book of Life" metaphoric network. But the metaphor is much wider and richer than this simple analogy. More generally, the orthodox perspective inherited from the Central Dogma continues to consider that if certain parts of DNA do not code for protein synthesis, it is because they have no function at all, that they are nothing more than the "vestiges of ancient information."[16] In fact, this first characterization of junk as fossilized information adheres to the most direct perspective of the original application of information theory. In this framework, single nucleotide mutation (SNM) is not only at the origins of evolution, but also at the origins of these nonfunctional vestiges: it is the source of noise. The linked trope appears regularly in the discourse on junk, as one example from the *New York Times* shows: "Biolinguists are trying to find a method for picking up the core three percent from the biochemical background noise and they are trying to spot the words without having to worry about what those words say."[17]

It is possible to view this perspective with Umberto Eco's code theory,[18] which improved the original model of information theory to introduce the s-code notion and so characterize the "genetic code" as a (c) type s-code,[19] "a receiver's set of behavioral responses." Eco extends the information theory's concept of "code" to four types of phenomena:

(a) A group of signals governed by internal combinatory laws, which he calls a syntaxic system;
(b) A group of states that are taken for a set of notions about these states, and that can become a set of communicable content, which he calls a semantic system;
(c) A group of possible behavioral responses on the part of the addressee, which he calls a pragmatic system;
(d) A rule that groups certain elements of system (a) with certain elements of system (b) or (c).

In equating the genetic code with a type (c) s-code, Eco thus considers the sequence of amino acids (in the cytoplasm) as a group of behavioral responses available to the addressee. But the DNA messenger can just as easily appear as a modulation of the heredity medium: the transcription of DNA into RNA by excising the noncoding parts (junk) and the permutation of T bases (Thymine) into U (Uracil) causes the signal to emerge through a type of modulation that may be qualified as digital (discrete). In this sense, Eco disregards that one can characterize the genetic code as a code in the full sense of the term (d), which is to say, as a rule that groups the syntactic system of transcription (a) to the pragmatic system (c) of translation of the mRNA codons into amino acid sequences.

However, it seems to me that this perspective has the distinct disadvantage of closing the mode of reference of the genetic code metaphor on its first reference: in metaphorically comparing DNA to the medium of heredity, this perspective winds up considering DNA as nothing but this medium. In other words, the description allowed by the metaphor becomes ontological. A second semiotic perspective will now allow me to reopen the genetic code's metaphoric mode of reference by avoiding the transformation of the first reference into ontology.

Indeed, if I pursue the semiotic description of DNA, I can just as well characterize junk from the group of insignificant details as carriers of an "*effet de réel*," as analyzed by Roland Barthes. From this view, the noncoding parts of DNA appear as "the irreducible residue of functional analysis": in other words, they appear as what resists the single meaning of DNA vis-à-vis an encoding

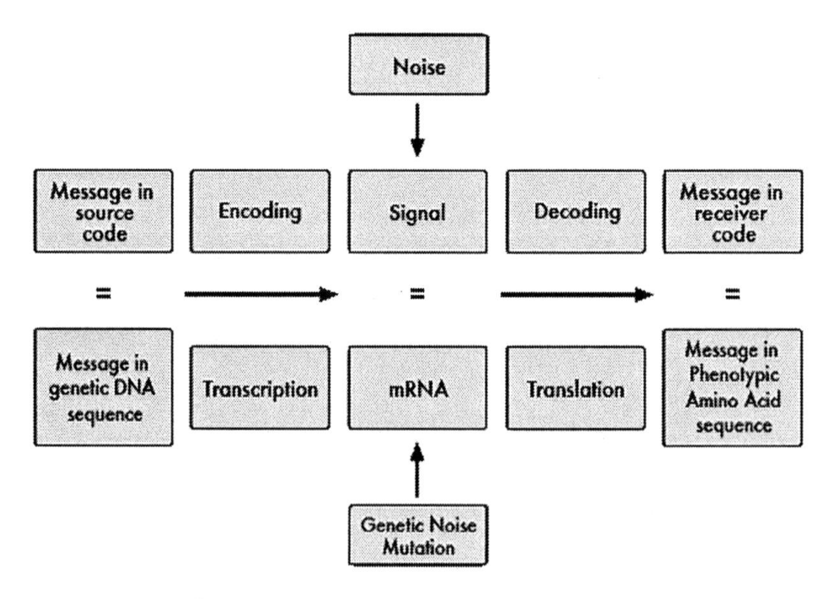

Figure 5. Genetic code equivalences.

function that is oriented toward protein synthesis. Their presence only confirms the orthodox account, both mechanistic and materialist, of standard molecular biology. As Barthes shows for text analysis, the presence of such details that are insignificant to the narration has no other effect than to render the narration more realistic: "it is the category of 'the real' (and not its contingent contents) that is thus signified; put otherwise, the very lack of the signified becomes the very signifier of realism: it produces *un effet de réel,* which is the foundation of a plausible yet unspoken truth that shapes the aesthetic of modern works."[20]

In this sense, the noncoding parts of DNA resist an ontological equation of DNA to the hereditary medium, a resistance that, as Barthes says, "confirms the wide, mythic opposition of real-life experience and of the intelligible."[21] Related to the semiotic view previously presented, such a perspective has the merit of recalling that the genetic code metaphor, in its primary mode of reference, consists of a story that is only more realistic insofar as a real part—of DNA—seems to escape it. Including details in the story that are insignificant to the narration thus paradoxically renders the story more realistic: thinking of noncoding sequences as noise makes a more realistic metaphor of a code that could organize DNA in terms of protein synthesis. DNA thus appears principally—and not uniquely—as the medium of heredity.

But once we concede this *effet de réel,* we must then consider the *effet de récit.* If insignificant details make the narration more "realistic," its meaning can also *produce* the real. In this inversion, I am following the flow described by Jean-Pierre Faye, from his *Hunic story*[22] up to *Murderous language,*[23] from the "sending off " of the former to the elliptic formulation of a "story effect" in the latter: "certain stories," he claims, "have *changed* the face or the shape of nations. The real history has been transformed by the way stories have been recounted and re-counted. To bring back the story is also to give the power of rapport: of making links and of their measure."[24] The former encodes both my thesis and its map as so many replies to a missing person (someone who is not there; an absent listener), like a receiver who is in search of a sender, or rather who might have learned to do without (a kind of modern weaning). The latter recapitulates them into a compact formula where everything is in italics. Placed here in an incessant backwards pivot, they finally allow my story to tilt into a second mode of reference, not abstractly declared, but instead directly told. All this for a face who can only see itself from a sideways glance, and who is laughing with the same sense of humor as the blind librarian from Buenos Aires, who provided me with a belated epigraph:

> More than once I shouted to the walls that it was impossible to decipher such a text. Imperceptibly, the concrete enigma that preoccupied me tormented me less than the generic enigma that is constituted by a sentence written by a god. "What sort of sentence," I asked myself, "could express absolute intelligence?" I realized that even in human languages there is no proposition that does not suppose the entire universe.[25]

The story of genetic code destroys the cybernetic metaphor's literal reference, while preserving it as the bootstrap program of an original story. Of this origin, this bootstrap, nothing remains but an endlessly reiterated command: *reboot!* From the Central Dogma of a one-way encoding, there remains nothing more than the ever-maintained possibility of (another) code in the code, an over-code, a meta-code, a supra-code, and so on. But the most orthodox developments of the original dogma have already taken back the inversion, while, admittedly, preserving its material function. Others have not been so conservative, and since 1994, many functions have indeed been imagined or demonstrated for noncoding DNA. These functions invite us to give a second wind to the metaphoric network of the "genetic code." The code metaphor's second reference loop can strengthen itself in the framework of a semiotic construct that re-presents junk DNA in the shape of another code, for the moment indecipherable: what if DNA were a multiplex medium, able to drive (or to transmit) both messages of protein synthesis and *other* messages?

This polyfunctionality, or this multimodality, allows us to imagine another language, or even "other worlds," with which DNA could put us in contact. In the era of the personal and distributed computer, this spun-out metaphor introduces meaning to the field of telecommunications, communication across time and distance, and even, why not . . . the quantum multiverse. Johnjoe McFadden's most recent thesis exemplifies just this, updating Schrödinger's original reference[26] by returning to the quantum intuition, to conclude that the 90 percent junk sequences of our own genome might be "capable of a kind of quantum sequence drift . . . invisible to the cell's quantum-measuring devices, allowing their mutational events to drift unnoticed into the quantum realm."[27]

In response to an e-mail in which I asked him about the origin of this notion of junk as a group of DNA sequences capable of quantum drift, McFadden laconically replied, "I'm afraid this is entirely (uninformed) speculation. My thesis is that DNA may drift into the quantum world whenever it becomes sufficiently isolated from the environment. Junk DNA seems to be the most likely to be isolated."

As we shall see now, we should thus consider not one but two basic narratives trying to make sense of genetic insignificance: the first, aka selfish DNA framework, became standard for a while; the second, however, was also available to me when I started this project. It is equated in my mind with one (very) peculiar book, and I will thus call it "the cosmic serpent framework."

OR3 **Bootstraps: Two Opposite Takes on Junk**

There is indeed one theory/metaphor about non-sense DNA that stands out since the late seventies, to the point that it renamed its object: Richard Dawkins's inspired *selfish DNA*. In *The Selfish Gene*, as early as 1976, and thus a couple of years before the "discovery of split genes," Dawkins proposed a "solution" for the apparent paradox posed by the "large fraction" of DNA that does not seem to "supervise the building of bodies": "The true 'purpose' of DNA is to survive, no more and no less. The simplest way to explain the surplus DNA is to suppose that it is a parasite, or at best a harmless but useless passenger, hitching a ride in the survival machines created by the other DNA."[28] A few years later, in his sequel, he insisted: "Nobody knows why the other 99 per cent is there. In a previous book I suggested that it might be parasitic, freeloading on the efforts of the 1 percent, a theory that has more recently been taken up by molecular biologists under the name of 'selfish DNA.'"[29]

Parasitic DNA, using its host, a mere material structure, our body, a survival machine. Freeloading DNA, taking the ride of its life, at the expense of protein machines (produced by the remaining 1 percent). Later in the book, Dawkins adds: "the genetic chromosomes are littered with old genetic text that is not used . . . these genes are either complete nonsense or they are outdated fossil genes."[30] Most importantly, he comes to this description via a computer metaphor: "to a computer programmer, the pattern of distribution of these 'genetic fossil' fragments is uncannily reminiscent of the pattern of text on the surface of an old disc that has been much used for editing text."[31] DNA is memory like a hard disk; coding DNA is the operative system of the computer.

Whose computer? Dawkins's very own, his *personal* computer: "a computer error (or, to be fair it may have been a human error) caused me accidentally to erase the disc containing chapter 3" (where Dawkins is both the cause and the perpetrator of the erasing). The rest of the story goes on explaining how he unerased his files, from "the bewildering jigsaw of textual fragments."[32]

We have here a new set of metaphors about junk DNA: computer memory overloaded with old textual fragments, evolutionary fossils, dormant parasites.

Dawkins's junk theory is definitely a cybertheory, and he is reinventing what cyberneticists have said for more than a century. Let me stress this point: if there is a metaphor here, it refers only to the computer, and not to "selfish." Dawkins insisted on this in his response to Midgley's (poor) critique of *The Selfish Gene:* to him, it is a literal, albeit astonishing, truth that "we are survival machines-robot vehicles blindly programmed to preserve the selfish molecules known as genes."[33] Indeed, Dawkins's selfish gene theory is grounded on a core cybernetic metaphor: in tune with the founding metaphor of the Artificial Intelligence (AI) research program,[34] genes are equated here with memory and computation, when the body's part is downplayed, reduced to a "false residual," or, as they put it in AI, a "meat machine."

Maybe because it was in tune with the original metaphoric configuration of molecular biology, Dawkins's theory received the approval of the pope of the discipline, Dr. Francis Crick himself. As we shall see next, on April 17, 1980, selfish DNA made the cover of *Nature.* In this issue, two review articles, and among them one cosigned by Crick, discussed Dawkins's ideas and turned them into the gospel on junk, inside the phenotypic paradigm: no function but survival, hence selfish. I will explain next how this gospel got to be.

The standard theory about junk DNA, aka selfish DNA, thus considers the body as a connected set of molecular machines. But when I started this project, I was aware of another theory, which looks at other kinds of molecular machines. In 1998, knowing my interest for the topic, one of my friends gave me a copy of a young anthropologist's book titled *The Cosmic Serpent: DNA and the Origins of Knowledge.*[35] When I read it, I realized that my whole project was taking a very different shape.

Trained at Stanford University with Terrence McKenna, Jeremy Narby lived between 1984 and 1986 with the Peruvian Amazonian Asháninka Indians. In the long tradition of anthropological studies of shamanism revamped by the ethnobotanical studies of the McKenna brothers, his work focused on the hallucinogenic knowledge of the Asháninka *ayahuasceros.* Facing the classic anthropologist dilemma—"Should we consider them alien or irrational?"—Narby took ayahuasca with his shaman "informers" and overcame his own materialist bias.

He then realized that there was a hypothesis that could answer to the puzzlement of Richard Evans Schultes, "one of the most renowned ethnobotanists of the twentieth century": "one wonders how people in primitive societies, with no knowledge of chemistry or physiology, ever hit a solution to the activation of an alkaloid by a monoamine oxidase inhibitor. Pure experimentation? Perhaps not."[36] The hypothesis required taking literally the *ayahuascero*'s answer to him when he asked why one sees snakes when one drinks ayahuasca: "it's because the mother of ayahuasca is a snake. As you can see, they have the same form."[37] So Narby took the next logical step and inverted the rational answer: DNA is the *cosmic snake,* and there are "possible links between the 'myths' of 'primitive peoples' and molecular biology . . . the double helix had symbolized the life principle for thousands of years around the world."[38]

Sharing my interest for metaphors and tautologies, Narby's first-person narrative is definitely on the speculative side. It provides an alternative hypothesis to Dawkins's standard neo-Darwinian theory that it amply criticizes. Quoting a microbiologist, Narby actually goes so far as to insinuate that his hypothesis is actually no more speculative than the standard "scientific explanation": "In fact, there are no detailed Darwinian accounts for the evolution of any fundamental biochemical or cellular system, only a variety of wishful speculations. It is remarkable that Darwinism is accepted as a satisfactory explanation for such a vast subject—evolution—with so little rigorous examination of how well its basic theses work in illuminating specific instances of biological adaptation or diversity." Narby's chapter 10, titled "Biology's Blind Spot," is to my mind the best summary of the vitalist argument of his theory of "minded DNA": "DNA corresponds to the animate essences that shamans say communicate with them and animate all life forms."[39]

At the beginning of my project, Dawkins and Narby thus provided me with two polar opposite ways to jump-start my inquiry. From this polarity stemmed the rest of my research. So, now that this background is set, let us forget the snake for a while (only to retrieve it later) and get back to the pages of *Nature.*

OR2 **The Selfish Contention**

On April 17, 1980, selfish DNA thus made the cover of *Nature.* In this issue, two review articles—signed by W. Ford Doolittle and Carmen Sapienza, on one hand, and Leslie E. Orgel and Francis H. C. Crick, on the other[40]—took up Dawkins's idea and reached the same conclusion: junk DNA is, quite simply put, "selfish DNA." Doolittle and Sapienza went even a notch further: "When a given DNA, or class of DNAs, of unproven phenotypic function can be shown to have evolved a strategy (such as transposition) which ensures its genomic survival, then no other explanation for its existence is necessary. The search for other explanations may prove, if not intellectually sterile, ultimately futile."[41] Survival thus became the "function" of this noncoding DNA and one could only marvel at the economy of such a "simple explanation." That was pretty much what Crick and Orgel did when they confronted Doolittle and Sapienza's proposition in the following fashion: after conceding that "it is an old idea that much DNA in higher organisms has no specific function," they insisted that "to regard much of this nonspecific DNA as selfish DNA is genuinely different from most earlier proposals," and they concluded with a customarily triumphant "the main facts are, at first sight, so odd that only a somewhat un-conventional idea is likely to explain them."[42]

The following issue of *Nature*, dated June 26, 1980, appeared with replies and criticisms from Thomas Cavalier-Smith, R. A. Reid, Gabriel Dover, and Temple F. Smith.[43] As expected, Cavalier-Smith criticized both papers for their (more or less explicit) position on the C-value paradox, and refused their "simple explanation." To him, they simply neglected "existing evidence for the idea that the overall amount of DNA in the genome has definite (nucleotypic), effects on cellular and organismal phenotypes, which are of profound adaptive significance."[44]

Gabriel Dover, on the other hand, criticized them from the junk-DNA perspective, but in the same direction. The two papers of the previous issues appeared to him as "throw[ing] out the baby (that small amount of suggestive evidence) with the bathwater and to replace it with genetically awkward concepts concerning the process of accumulation of 'junk.'" Moreover, Dover claimed that there was still a need to underline "the limitless ways of creating and disseminating 'junk' by random process" rather than "relying on the more limited idea of the genome as an arena for selfish replicators under 'non-phenotypic selection.'"[45]

Until then, the argument was quite predictable: Doolittle and Sapienza and Orgel and Crick transposed Dawkins's notion of selfish genes to make sense of the noncoding sequences of DNA, and therefore exposed themselves to criticisms from both the proponents of the adaptive hypothesis (represented by Cavalier-Smith) and the junk-DNA hypothesis (represented by Dover). The next move, although hardly surprising at a strategic and conceptual level, was still somewhat of a surprise: *Nature* issue 288, dated December 18–25, 1980, appeared with final reply from Orgel, Crick, and Sapienza and Dover and Doolittle.[46] The original authorship was reorganized and the most agreeable critic was co-opted. Orgel, Crick, and Sapienza, three of the original four authors, restated the original points with more semantic nuances (see below), and answered Cavalier-Smith. They cleared up a misunderstanding they had had with Cavalier-Smith's proposal, but they also repeated that they "did not find all [his] assumptions particularly plausible." They concluded:

> In our recent experience most people will agree, after discussion, that ignorant DNA, parasitic DNA, symbiotic DNA (that is, parasitic DNA which has become

useful to the organism) and "dead" DNA of one sort or another are all likely to be present in the chromosomes of higher organisms. Where people differ is in their estimates of the relative amounts. We feel that this can only be decided by experiment.[47]

Dover and Doolittle, on the other hand, produced a middle ground between the junk and the selfish DNA hypotheses, thus creating the basis for a reconciliation of the two variations of the same argument (hard = selfish, soft = ignorant). They did not address Cavalier-Smith's position or the adaptive hypothesis. Instead, they reconfigured their vocabulary as follows: "we agree that the amplification and dispersion of segments may occur either at random (sequence independent or 'ignorant') or with preference for certain sequences (sequence dependent or 'selfish')."[48] So it was quite a different point that they made when they too concluded that more experimental work was needed to decide between "different alternatives": the alternatives they had in mind were not at all between those provided by the adaptive and the junk hypotheses, but only between two variations of the latter, ignorant and selfish DNA. "The problem now, as with most scientific debates," they concluded again, "is one of quantification."[49]

One could say that the call for more experimental work and quantification was a very diplomatic (not to say shrewd) way to close the argument. The original claims were reduced to competing hypotheses, and the strange structure of the reorganized authorship of the final replies allowed them to both answer the most destructive criticism and restate, albeit in a nuanced fashion, the junk hypothesis, now provided with two variations.

While attempting to understand the issues at stake and the closure of the argument, I was struck by the number of references to the semantic question in the papers published in *Nature.* For instance, Orgel, Crick, and Sapienza started their final reply by stating that "certain difficulties have been caused by the words 'selfish,' 'junk,' 'specific' and 'phenotype' that were used in the two reviews of selfish DNA."[50] They then discussed in detail their own take on the terminology. In order to get behind the scene on this point, I turned to Francis Crick's correspondence between August and November 1980.[51] Before I get back to the semantics issue, let me first summarize how this correspondence confirms the strategic reading of the closure of the argument that I just provided.

The correspondence shows precisely the process of co-optation of Gabriel Dover. It reveals that it started with a meeting between Doolittle and Dover: "Gabriel Dover and I met in Vancouver this summer and realized that the differences in our opinions were not fundamental" (FDtoPN). Doolittle then told his colleagues that the disagreement between them and Dover was not that important, and that "Dover informed [him] that he would probably agree with whatever it was we had to say about Cavalier-Smith" (FDtoLO).

Dover then exchanged letters with Crick and they easily reached an agreement. In his letter, Dover insisted that his criticism wasn't "meant to be polemical," and that he was "in very broad agreement with you on most of this" (GDtoFC). The following statement from Doolittle's letter to Crick confirms the proximity between the two versions of the "junk hypothesis": "the points that Dover raised in his 23 October letter to you, although probably in most cases valid, are not tremendously important" (FDtoFC). In the end, the correspondence shows that a consensus was easily built between Dover and the original four authors, at the expense of the criticisms made by Cavalier-Smith. Dover even wrote to Crick that he suspected that Cavalier-Smith "was simply trying to score a point by historically pre-empting your use of the terms parasitic and selfish" (GDtoFC).

In Dover's strategic reading, the semantic question therefore takes a central part. In their correspondence, however, our five authors regularly express their disinterest for the semantic aspect of the question debated during the controversy: "a waste of time and space" (LOtoFC); "fruitless and repetitive semantic argumentation might thus be aborted" (FDtoPN).

In fact, they seem to consider that the semantic question was secondary compared to the point that they wanted to make. In an intermediary paper prepared in the process of making the final reply, Doolittle and Sapienza made this point clearly: "We do *not* wish to argue about what this kind of selection should be called, or by what adjectives we should describe its products; it is the recognition that this kind of selection logically must have and demonstrably has occurred that is important' (Ms. 0). Ms. 1.1 also includes a passage deleted in the final manuscript that makes the same point: "there is a problem of language but it would be a pity to lose the opportunity to summarize these processes because we cannot assign acceptable terms to them." In his letter to Francis Crick, Ford Doolittle gave the final argument about the whole semantic question: the idea is more important than the terminology, he contended, and even added that "a profusion of terms is maybe not such a bad thing; it allows others to choose among a variety of ways of describing the same phenomenon that which is least offensive to their biopolitical sensibilities" (FDtoFC).

Many other excerpts from the correspondence confirm this notion of a determining relationship between semantics and "biopolitical sensibility." For instance, Francis Crick wrote to Ford Doolittle: "I found that, in my travels that many people accepted the idea if this term ['parasitic'] were used but baulked when I said 'Selfish DNA,' so I feel people should have the chance to decide for themselves" (FCtoFD).

The discussion carried out in the correspondence about this specific term ("parasitic") seems important because it was the alternative word that Crick and Orgel, after Dawkins, had chosen to characterize "selfish DNA" in their title: "the ultimate parasite." Dover clearly had both a logical and a "biopolitical" problem with this choice of word:

> I dislike the word "parasitic" because it has, in popular usage two connotations that do not apply. One is horizontal infection and the other that it is harmful. In fact, the second is the more serious problem on the basis that it is being proposed that this DNA has no large effect on fitness, and accumulates more in accordance with its own self-perpetuating properties. The word parasitic strongly implies the harmful exploitation of another organism. (GDtoFC)

Crick answered that they had indeed "intended 'parasitic' to imply (in its pure form) some disadvantage to the host" (FCtoGD). But he also conceded, once Doolittle had signified his agreement with Dover, that he was not "wild" about it either (FCtoFD). In their final reply, however, Orgel, Crick, and Sapienza kept using the word *parasitic*, while Dover and Doolittle did not. This might explain why the original project of having only one final reply signed by the five authors (FDtoLO) did not happen. My guess is that it is a possible interpretation, because the consensus between the two versions of the "junk hypothesis" (selfish and ignorant) does not require another negative characterization (it is already assumed in all the terms), when the opposition between the junk hypothesis and the adaptive hypothesis (Cavalier-Smith's position) is based on it.

The following exchange of letters between Dover and Crick can be seen to make this point in recapitulating all the potential terms to be used. Dover starts by stating his position that "no new word over and above repetitive DNA" was needed. He added that "Non-coding DNA is an alternative, and that 'Junk' is O.K. if defined in contrast to specific" (GDtoFC). Crick answered "we also feel 'non-specific' does not convey our meaning," but he added that "'repetitive' does to some extent but has been blunted by use." He also insisted: "We only use 'junk' in that paragraph and say that we hope better terms will be devised" (FCtoGD). In the end, the following excerpt from Dover's letter seems to summarize quite well the issue, but left the semantic question unsolved: "Wouldn't it be better to simply say (without definitions) that we are dealing with sequences that can accumulate intragenomically because they appear to be non-specific in their effect but were they to affect the genes then their effects become specific and accumulation is restrained?" (GDtoFC).

OR1 **Even the Sharpest Razor Cannot Shave Its Handle**

Thus one was left with the paradoxical impression that semantics were both crucial and unimportant in the argument so far. Eventually, the final replies seemed to leave this question unanswered but blatantly kept using the same terminology that had proved so controversial. More important, this was also the very point of a criticism that did not get answered in the final replies. I will now attempt to demonstrate that this omission was highly relevant to our understanding of what was discussed then.

None of the replies addressed the comments made by Reid in the June issue, because, indeed, it was more a commentary than a critique of the original papers. Reid proposed that certain mutants of *Sacharomyces cerevisae* were "promising systems for investigating the questions raised by Orgel and Crick."[52] However, Smith's paper was treated in the same fashion—only one of his most technical points was replied to by Orgel, Crick, and Sapienza—when he offered a couple of other criticisms that they chose to ignore. I now want to get back to these criticisms, because I think that they provide a logical point necessary to understand the closure of the argument. Coming back to the authors' contention that "since the evolution of properties such as transposability ensure sequence survival, no other selective or functional properties are required to explain the existence of much of this 'extra DNA,'" Smith rightly pointed out that "these arguments are at their base a variant on Occam's razor."[53]

When his point seemed to confirm both replies in their conclusions that more experimental work was needed, it also put this conclusion in a context worth noting. Pointing to the core logical argument of Occam's razor, Smith alluded to the nominalist background of modern science. Charles Sanders Peirce once noted that "this brocard, *Entia non sunt multiplicanda praeter necessitatem*, that is, a hypothesis ought not to introduce complications non requisite to explain the facts, this is not distinctively nominalistic; it is the very roadbed of science."[54] By this he meant that the razor might have come from the nominalist tradition, but, more important, it became the crucial criterion of scientific methodology. Karl Popper in *The Open Society and Its Enemies* agreed and argued convincingly that the natural sciences are methodologically nominalist. Smith's point thus appears to be historically valid. However, it is worth reminding oneself that Occam's razor is "powerless for actually *denying* the existence of certain kinds of entities; all it does is prevent our positively affirming their existence."[55] In this crucial reminder, Spade summarizes two key methodological statements: (1) the razor does not guarantee that one has got all the genuine entities needed for the explication, and (2) a more logically correct statement of the principle by Occam himself is that "when a proposition is verified of things, more [things] are superfluous if fewer suffice."[56]

Now, Smith's point that selfish arguments "are at their base a variant on Occam's razor" can be made clearer in the following fashion: if the selfish DNA proposition is that noncoding parts of DNA can be proved to result from an intragenomic mode of evolution (that has no effect on the phenotype), then there is no need to try to find them any other function. Or perhaps even more clearly: although this argument cannot allow us to deny the existence of such a function, it prevents us from positively affirming its existence. The logical problem arises when one realizes that the selfish DNA proposition was at the time of the argument (and still is today) a *hypothesis:* in other words, nothing has yet proved that *all* noncoding DNA indeed results from such an "intragenomic mode of evolution": the evidence was (and still is), to say the least, fragmentary. Smith understood this, and pushed his criticism further, onto the epistemological level: he claimed—and I believe rightly so—that the selfish DNA theory "appears nearly irrefutable":

> If a phenotypic constraint or function is found for any given sequence, it is either removed from consideration under the theory or it is argued that its function was a later adaptation exploiting the already existing parasitic sequence. Similar considerations have, of course, plagued the theory of evolution from its inception.[57]

In his later review of the Orgel and Crick paper, Stephen Jay Gould noted the same problem: "the confusion of current utility with reasons for past historical origin is a logical trap that has plagued evolutionary thinking from the start."[58] But, moreover, he pointed to the true nominalist problem raised in the controversy: what is (are) the appropriate unit(s) of selection?

> If bodies are the only "individuals" that count in evolution, then selfish DNA is unsatisfying because it does nothing for bodies and can only be seen as random with respect to bodies. But why should bodies occupy such a central and privileged position in evolutionary theory? To be sure, selection can only work on discrete individuals with inherited continuity from ancestor to descendant. But are bodies the only kind of legitimate individuals in biology? Might there not be a hierarchy of individuals, with legitimate categories both above and below bodies: genes below, species above?[59]

One only has to change "individuals" for "singulars" to understand the nominalist origin of the problem . . . into the old "quarrel of the universals"![60] It was indeed an extreme nominalist notion to consider that only singulars are real entities, but how could that apply to genes? To answer this, I am afraid, would require another chapter.[61]

The argument studied in this chapter appears to be a crucial turning point in the history of the genetic code (and genetic insignificance): what was at stake is not the further enabling of the construction of an experimental world based on the phenotype paradigm, but its restriction. One could speak here of negative performativity, closing of the discursive field, exclusion of potential rhetorical and experimental practices. I contend here that the *Nature* episode marks the inversion point of the phenotype paradigm's rhetorical program.

At the closure of the argument, Orgel, Crick, and Sapienza included the word *phenotype* among the (dirty) words that had "caused difficulties." Like the other incriminated words (*selfish, junk,* and *specific*), they then devoted a paragraph to dispelling the semantic ambiguity by proposing two words: "one to refer to the phenotype of the organism and the other to apply solely to the 'phenotype' of the parasitic DNA, a distinction we would certainly make in the case of a true parasite."[62] For the former they proposed "organismal phenotype," and for the latter, following Cavalier-Smith, they argued for "intragenomic phenotype." But the reference to Cavalier-Smith is misleading here, because in his criticism this author never used the expression "intragenomic phenotype." Instead, he wrote about "intragenomic *selection,*" and contended that it was clearer than "non-phenotypic selection," because, according to him, "it directs attention to the fundamental phenomenon in question—intragenomic competition between different sequences—without begging the question as to the phenotypic effects of the competing sequences, or misleadingly implying that there is a well-defined constraining concept of phenotypic selection or 'phenotypic benefit.'" He even concluded that "since this new restricted use of 'phenotype' to exclude certain traits (but which?) associated very directly with DNA has never been clearly defined, the concepts of 'non-phenotypic selection' and 'selfish DNA' themselves lack precision."[63]

Thus, Cavalier-Smith rather seems to advocate here the exclusion from the vocabulary of the very concepts that Orgel, Crick, and Sapienza proposed: there can be no "intragenomic phenotype," no more than there can be "organismal phenotype" (his "restricted use"). What we are left with is a profound problem with the notion of phenotypic effect as the sole criterion for selection and a new mode of selection, intragenomic, which cannot be explained with any reference to a "phenotype."

Any further attempt to exclusively use the criterion of phenotypic effect to explain evolution is doomed: Orgel, Crick, and Sapienza's attempt to artificially maintain it through a "context dependent" disjunctive terminology is the desperate attempt to inscribe, and therefore hide, the tautology in the terminology. It only avoids begging the question *in appearance.* To follow their lead would turn "phenotype" into a metaphor. This would be the absolute metaphoric breakdown of the original phenotype paradigm of molecular biology as we knew it. Such is the conclusion of Gould's review of Dawkins's version of the same idea, under the name of the "extended phenotype":[64]

> I have always admired the chutzpah of Senator Aikens' brilliant solution to the morass of our involvement in the Vietnamese War. At the height of our reverses and misfortunes, he advised that we should simply declare victory and get out. Richard Dawkins got in with his 1976 book, *The Selfish Gene.* He declared victory with *The Extended Phenotype* in 1982—although he had really, at least with respect to the needs and logic of his original argument, gotten out.[65]

What better metaphor to spell out the misfortune of an unperformative metaphor (one that Gould calls, ironically, recycling Dawkins's terminology, an "impotent meme")?

Far from being a pejorative disqualifier, to define as "semantics" the crux of the argument studied here puts the emphasis on one central tenet of modern science: it can be constrained as well as enabled by its choice of words. This, indeed, is merely a truism in respect to its nominalist origin, and certainly no news for contemporary philosophers of science.

Steven Shapin has argued convincingly that the scientific revolution was first an affair of *gentlemen,* "substantially defined by [their] reliability, [their] promise, [their] word."[66] Quoting Shakespeare, in the inverted plural this time: "his words are bonds, his oaths are oracles." From this play on word(s), the historical motto of the Royal Society, *nullius in verba,* appears quite paradoxical: "the experimental credo professed to rely upon 'no man's word' and to accept the testimony of nature itself. In fact such a prescription was, and is, impossible to act upon. The prescription is best taken as a normative disengagement from certain institutionalized practices and sources of authority in favor of others. Some spokesmen for reality were replaced by others."[67]

This shift in authority (rather than the destruction of authority) is usually described as a shift from the authority of schoolmen, convoking multiple authorities to discuss endlessly the sex of the angels, to the authority of scientists, ultimately derived from direct experience of nature. I hope that this chapter will have convinced you, dear reader, that today's scientific achievement can also rely on authority, metaphor, and speculation. However, whether this is a sound conceptual diagnosis is of little interest *in abstracto*, but rather a pragmatic point worth making. In the manner of the only trustworthy journalistic practices, I offer here two confirmations that will show that some practicing scientists indeed seem to think so.

During my interviews for this book, I had the opportunity to interview two very different scientists on this question.

The first one, Dr. Roy J. Britten, can be described appropriately as the pioneer of the now numerous studies demonstrating that there are deep regulatory functions buried in so-called junk DNA (we shall meet Roy Britten again in chapter 2).[68] He definitely inscribes his work inside the dominant paradigm in molecular biology. During our interview, he offered this laconic statement about the two original review essays that started the controversy: "They were both pretty bad papers. And because of his [Crick's] position of course . . ." He then gave me a classic Kuhnian explanation of the whole thing when I asked him if it was risky for a career to study repetitive or noncoding DNA:

> RB: I suppose. There was a long period when I had it pretty much to myself . . . Most molecular biologists wouldn't do it because they were ignorant of that whole field until [Barbara] McClintock won the Nobel Prize [in 1983]. That's all a funny history . . . At that time, geneticists just ignored it, because it was outside the paradigm in which genetics work, so forget it! It wasn't ignorance; they would claim, you know, she wrote complicated papers or anything. That's all baloney. They knew exactly what she had written.
>
> TB: In which way exactly was it outside the paradigm? You say it was outside of the paradigm, what made it outside?
>
> RB: Because [she held that] genes were moving around and involved in the control of other genes. That transposition was important in control. And that is, well in fact, they were not paying much attention to control at all.[69]

He concluded our interview on a tremendous "now you know what junk DNA is about."[70]

To debunk Britten's narrative would require going back to McClintock's work and understanding how it was framed for too long outside of the dominant paradigm. Evelyn Fox Keller has offered a starting point to do that with her remarkable biography of Barbara McClintock.[71] Rather than focusing, in a Kuhnian fashion, on paradigms and scientific communities, Keller chose to focus on the individual, on the "idiosyncrasies of autobiography and personality."[72] At the end of her book, however, she left the last word to Barbara McClintock herself: "we are in the midst of a major revolution that will reorganize the way we look at things, the way we do research . . . and I can't wait. Because I think it's going to be marvelous, simply marvelous."[73]

The second scientist I interviewed, Dr. Colm Kelleher, has long felt the inertia of this revolution in the making. A younger scientist, he studied and practiced for fifteen years standard molecular biology with all the signs of a well-behaved career. Unexpectedly, a few years ago, he answered successfully an ad in *Science* to join the National Institute for Discovery Science, a private effort to study so-called anomalies (in Jacques Vallée's[74] rather than in Thomas Kuhn's sense). He then became what I can only describe as "Las Vegas's very own real-life Fox Mulder" (we shall meet Dr. Kelleher again in chapter 3)—not exactly a "normal science" typical foot soldier. Here are his recollections about the impact of the argument presented here, while he was working then at the University of British Columbia in Canada on retrotransposons:

> I know that those papers were very influential in killing the idea of granting, making grants available for work on the noncoding DNA. So there is a kind of lineage there which, I suppose, had an impact on the field, but there is [also] an emotional thing there. Because that whole viewpoint that junk DNA is junk DNA, therefore it's garbage, therefore it's not worth studying, therefore it's not worth funding people to study it, directly impacted me.[75]

After all, which scientist in his or her right mind would claim to practice *futile* research? Fortunately enough, even the advocates of the standard paradigm in molecular biology know now that junk is no garbage. Sydney Brenner, for instance, signed a commentary of a 1997 paper where Stephen Jay Gould was deploring the pejorative connotations of "junk" to refer to the noncoding parts of DNA. Brenner wrote: "some years ago I noticed that there are two kinds or [*sic*] rubbish in the world and that most languages have different words to distinguish them. There is the rubbish we keep, which is junk, and the rubbish we throw away, which is garbage."[76] Brenner added: "The excess DNA in our genomes is junk, and it is there because it is harmless, as well as being useless, and because the molecular processes generating extra DNA outpaced those getting rid of it." He seemed pretty affirmative there, and certainly Britten as well as Kelleher would not agree with him on such a definitive statement of causality. Rather, they would argue: "Garbage we throw away, indeed, but junk we keep . . . just in case it might be useful." The search for a function for junk DNA (or at least for some of its parts), even if it was once declared with some authority *if not intellectually sterile, ultimately futile,* goes on.

It appears thus quite ironic, retrospectively, that the acme of today's big science, some three centuries after the modern scientific revolution, ends up producing a controversy where authority over metaphor is crucial. It seems even more ironic to me that the methodological principle put forth during the controversy, the famous razor, was indeed the last legacy the schoolmen left to their modern inheritors.

cro **Genes and Signs of Meaning**

I started the background section of this chapter by stating the essential "ambivalence"[77] of the notion of genome, as both the full set of chromosomes (whole DNA) and all the genes contained therein. It now seems that this first ambivalence is actually mirrored today in the notion of gene itself, as both "any sequence of DNA" and "a *coding* sequence of DNA."

"Gene" was first the name given by the Danish geneticist Wilhelm Johannsen to a *hypothetical* entity—others before him had called it "a living particle" (Hagedoorn), "a minute granule" (de Vries), or "*elemente*" (Mendel)—which he felt was needed to account for the transmission of hereditary characteristics (intergenerational transmission). It soon also became the name of the agent responsible for the formation of traits or characteristics of a given organism (intragenerational development). Evelyn Fox Keller has summarized this by calling the gene "a monster . . . [that] offers a resolution to the riddle of life by invoking an entity that is a riddle in and of itself."[78]

Lenny Moss has clearly restated this dual origin in his history of the gene concept, which leads him to distinguish two notions of the gene, unfortunately often "conflated": "one the heir to performationism and the other the heir of epigenesis. 'Gene-P,' the performationist gene concept, serves as an instrumental predictor of phenotypic outcomes, whereas 'Gene-D,' the gene of epigenesis, is a developmental resource that specifies possible amino acid sequences for proteins."[79]

These points are now well established. But there might be yet another duality, or another "conflation," hidden in many uses of the word *gene* in twentieth-century biology. Moss argues convincingly that "the conflation of two individually warranted but mutually incompatible conceptions of the gene . . . were held together by the rhetorical glue of the gene-as-text-metaphor."[80] The discovery of the structure of DNA, followed by that of the genetic code, indeed could only work on the basis of this metaphor, albeit in destroying the original, cybernetic, literal sense of "code. " I have argued elsewhere that this destruction enabled the construction of a new sense, new experimental practices that ended up extremely successful for a renewed understanding of the mechanisms of heredity.[81] Maybe too successful!

In the process, "structural genes" and "regulatory genes" were invented, and everything went peachy for a while. Until, that is, the "good shake" given to these ideas by the discovery of splicing, this "mini-revolution in molecular genetics" (Crick). Remember, the gene was first a postulate, a hypothetical entity. Once the structure of DNA and the genetic "code" were established, it was thought that DNA was, simply put, a collection of such genes. Or, inversely, that genes were bits of DNA, and that *DNA was solely made of such bits.* The discoveries of the structure and the code of DNA lead one to believe that genes were not a hypothesis anymore. They had acquired the only existence scientists seem to believe in, physical, that is, material existence. In a truly nominalist fashion, genes became real once they were given a physical presence (before that they were mere "names," i.e., fictions of the mind).

"Once upon a time," writes Moss, ". . . it was the happy time of biomolecular simplicity, summed up in one equation: DNA (= genome) = complete set of genes." According to him, "beginning with the discovery that eukaryotic genes are assemblages of ancient modules (Gilbert 1978) and with the recognition of the actual dynamism of DNA, a very different picture has progressively emerged."[82] While this statement is historically valid, and the "very different picture" indeed started to emerge after the discovery of introns (1977), it hides a more perplexing truth. There *was* trouble in the house of molecular biology prior to 1977 and the dynamism of DNA was explored long before that time. Britten and Ohno at first, and many other scientists after them, knew since the early 1970s that there was trouble, and Barbara McClintock had pretty much focused on the "dynamism of DNA" in her work since the mid-1940s. The C-value paradox was actively studied. So, why did it take the discovery of introns to start the "minirevolution"?

It took the discovery that genes were not "linear things" to realize that the very notion of a gene was in danger. In my mind, the controversy studied here brings forth an interpretation of how exactly the danger was avoided. The foreclosure of the epistemic field it allowed, reversing the performativity of the cybernetic metaphor through an improper use of Occam's razor, put out the fire (at least for a while). There were many more gains to be harvested from "the old picture," by concentrating on the coding part of DNA. Crick et al.'s decision that junk was selfish, moving from selfish *gene* to selfish *DNA*, was actually trying to *rescue* the phenotype paradigm.

There is this part of DNA, you see, that we don't know how to name. To call it by a given name might have consequences on the amount of time, energy, and funds that are spent in research. All those metaphors/names have performative aspects (and reflect different "biopolitical sensibilities"). To call it "selfish" or "junk" amounts to denying any (or little) need for such investments in this line of research. In order to maintain molecular biology's supremacy over the field of genetics and development (and expand it to the field of consciousness), researchers had to stick to the gene-as-text metaphor and find in its semantic realm an adequate name for noncoding DNA. "Selfish" did the job, precisely because "Dawkins's selfish replicator constitutes the quintessence of conflationary confusion."[83]

If one wants to continue to consider the whole DNA as genome, one would have to complete the dualism of Gene-P and Gene-D by another kind of "gene." Certainly not Gene-S, if -S stands for Selfish! Although S might also stand for *Sign.* In which case, as Hoffmeyer and Emmeche have proposed, we only need S-Genes: "Genes must be understood as signs, not as particles or 'pieces of DNA.'"[84]

The S-Gene theory, which in fact characterizes the principal tenet of the field of biosemiotics, rests on an open metaphor. The main point of this chapter is that metaphor is a crucial resource of the scientific enterprise when its mode of reference is open. In this case, the performativity of the metaphor is positive, that is, it can actually produce an experimental world and build its own referent. The breakdown of the original cybernetic metaphor of molecular biology that I diagnosed in studying this episode in *Nature* means just the opposite: it closes the metaphor and puts forth a negative performativity. Now, the metaphor prohibits certain referents and constrains explications inside a certain semantic (and therefore epistemological) realm. Because of a logically flawed, authoritative decision, genes are to remain material entities with only one possible "meaning," that is, only one interpretant, that of protein synthesis. The quest for their interpretants is nevertheless not over, and the list of their effects on an organism remains open.

N **Why Junk Might Remain Selfish**

If research on junk DNA was discouraged, if not prohibited, in 1980 by the highest authorities (i.e., Francis Crick and his colleagues in *Nature*), it was for a while only. Since the early 1990s, the ban was lifted, at least to a certain extent. But some consequences last to this day: the only theses well accepted on this innocuous object are still those actually confirming the standard (i.e., neo-Darwinian) view. If there were, after all, some functions to be found for some parts of noncoding DNA, they should only make sense with respect to the coding functions: mostly regulatory or structural. One of the results of this state of affairs is that, for the general public—outside of the scientific community and some margins of the educated public—the standard model of molecular biology, and even the Central Dogma, remain unchallenged.

Case in point: on the World Wide Web—after all, one of today's main repositories of current folk knowledge—the best indications of this state of affairs are to be found on wikis ("the simplest online database that could possibly work," i.e., the current implementation of a collective Web trail), and especially on *Wikipedia* and other related Web encyclopedia in a wiki form. Here are two entries on "Junk DNA" from two such wikis:

> In molecular biology, "junk" DNA is a collective label for the portions of the DNA sequence of a chromosome or a genome for which no function has been identified. About 97% of the human genome has been designated as junk, including most sequences within introns and most intergenic DNA. While much of this sequence is probably an evolutionary artifact that serves no present-day purpose, some of it may function in ways that are not currently understood. Recent studies have, in fact, suggested functions for certain portions of what has been called junk DNA. The "junk" label is therefore recognized as something of a misnomer, and many would prefer the more neutral term "non-coding DNA."[85]

> The genome of almost all eukaryotic organisms contains a large proportion of DNA which does not code for proteins. Some of this DNA is concerned with regulation of genes, but much of it has no discernable purpose and is therefore popularly known as "junk DNA" . . . There are a number of different classes of non-coding DNA which forms part of genes, notably regulatory elements and introns . . . However, introns do not take up much more of the genome than coding DNA does, and so do not provide a general function for "junk DNA" or the C-value paradox.[86]

Although I characterized wikis as current repositories of "folk knowledge," it is obvious that the editors of the quoted entries on "junk DNA" belong to the scientifically educated public. In fact, the main editors of the first entry are graduate students from some of the best institutions in the United States (e.g., MIT). One of them, a postdoc in the Molecular, Cellular and Developmental Biology Department at Yale University, even carries out research on noncoding RNAs. It is even clearer that the editors of the second entry share an interest for the "biopolitical" aspect of the question: one of the three main editors introduces himself as "a physicist born in 1963 [who] doesn't like pseudo science," and another one gives the following self-presentation: "a biology nut with a spatial/geographical perspective." But he also insists that he tries "to bring the biology back into the center of the [Intelligent Design] debate, where it should be." This shared interest is quite obvious in the introduction of their entry on "junk DNA," where they claim that "creationists feel uneasy about having a genome full of DNA that doesn't do anything—why would God be so wasteful?—and claim that junk DNA isn't junk DNA."[87]

The statement "junk DNA is not really junk" thus became "an antievolutionist claim." Here is why junk DNA might remain selfish: the biopolitical aspect of the question is even more loaded in 2005 than it was in 1980. Junk DNA has become a major bone of contention in the renewed controversy between (neo)creationists and (neo)Darwinians in the United States.[88] And they fight to the bitter end, as Richard van Sternberg has learned to his cost.

Richard van Sternberg's story, indeed, shows that the question has become an ideological issue (and I don't mean this in a good way). Holder of two PhDs (one in molecular evolution [1995] and the other in systems theory and theoretical biology [1998]) and a research associate at the Smithsonian Museum of Natural History since his postdoc in 1999, van Sternberg might remain in history books because he is the first managing editor to publish a paper about intelligent design in a peer-reviewed journal. *Woe to that man by whom the offence cometh!* (Matthew 18:7).

On August 4, 2004, van Sternberg published an extensive review essay by Dr. Stephen C. Meyer in the *Proceedings of the Biological Society of Washington* (117[2]: 213–39). Dr. Meyer was at the time, and still is, the director of Discovery Institute's Center for Science and Culture, perhaps the major private institute backing the intelligent design claims. In this paper, Meyer argued, "no current materialistic theory of evolution can account for the origin of the information necessary to build novel animal forms." He proposes intelligent design as an alternative explanation for the origin of biological information and the higher taxa.[89]

A fierce controversy ensued—the paper was debated at length on the Web in forums and newsgroups, the peer-review process was questioned, and van Sternberg was eventually threatened from inside both the museum and the NIH, his employer. The issue made the headlines of the national press in late August 2005 (*Washington Post, New York Times,* etc.), and was still raging when I first wrote these lines, in 2007. Expertise and counterexpertise were mobilized, but the damage was done: a first peer-reviewed paper on intelligent design had been published.[90] One has to feel that the whole process was quite timely for President Bush's agenda. After all, intelligent design is but another name for neocreationism, and the U.S. president had created quite an uproar himself on August 1, 2005, when he "waded into the debate over evolution and 'intelligent design' . . . saying schools should teach both theories on the creation and complexity of life."[91] Yes, indeed, the issue had become an ideological one. And junk might thus remain selfish . . . for ideological reasons.

Q Design

Obviously, the creationist/evolutionist controversy did not start in August 2005. President Reagan, before President G. W. Bush, had already endorsed the position that creationism should be taught in the nation's schools.

Moreover, it seems that "the controversy" spanned most of the twentieth century in the United States.[92] And, in fact, the issue at stake here even predates Darwin and his theory of evolution, and seems to be . . . well, essentially human. One could say, following Whitehead again, that the evidence of design in nature (i.e., the design argument for a creationist perspective) appears as an "eternal object" to human eyes (or mind, or even soul). Many writers in history have ascribed this notion to the core of the animist belief, and tracked it back to our earliest ancestors. This "caveman story" rests on an anthropocentric argument that has been rejected by many critics of the design argument, from Hume to Spinoza . . . and, eventually, Richard Dawkins.[93] Other writers have found in pre-Socratic philosophy the premises of such an argument.[94]

Most writers hold that the design argument found its fullest premodern expression with Saint Thomas Aquinas's (1225–1274) fifth way to prove the existence of God on the basis of what can be known from the world. Also named the teleological argument, Aquinas's fifth way is a theologian's answer to the rediscovery of the Aristotelian corpus, or, more precisely, "a fundamentally non-Aristotelian correction of Aristotle in a corpus that is usually considered to be a synthesis of Christian faith with Aristotelianism."[95]

This fifth way became the basis of what is known as "natural theology" in the early-modern times. *Natural theology* is the attempt to derive knowledge of God and of his attributes solely on the basis of reason and experience (in contradistinction to *revealed theology*, derived from scriptures). Even if it seems hard to imagine by looking at most present-day scientists, experimental science was not, at first, necessarily atheist. Most heroes of the "scientific revolution" were Christians, and experimental science was still strongly rooted in the Christian worldview.[96] What was new, and truly revolutionary, was to entertain the idea that God could be known via the Book of the Bible *and* the Book of Nature,[97] and that the latter was written in the language of mathematics rather than that of metaphysics.

Early scientists indeed attempted to understand the mathematical language in which the world was written *for the Glory of God,* to convince atheists and unbelievers of his existence by describing the intricacies and laws of Nature. In this process, they often used images, analogies, and metaphors (contrary to what they did when addressing one another, as prescribed by the early reform of prose and the "naked style"). Robert Boyle, one of the adamant proponents of these new literary technologies, "was responsible for developing an early analogy between the universe and a clock,"[98] which Shapin and Schaffer call "the root metaphor of the mechanical philosophy."[99]

Natural theologies of the eighteenth and even nineteenth centuries took over this root metaphor and translated Saint Thomas's fifth way in its terms: the analogy of the watch and the watchmaker, this new version of the design argument, was born. Its first enunciation might have come from Bernard Nieuwentijdt (1654–1718), a physician and burgomaster in the small town of Purmerend, in his *Het regt gebruik der werelt beschouwingen, ter overtuiginge van ongodisten en ongelovigen* (1716), which argued against Spinoza. It went through several editions and was translated into English as *The Existence of God, Shown by the Wonders of Nature.* Voltaire owned a copy of this book, and it was an influence on William Paley, author of one of the most famous versions of the analogy.[100] In his book, William Paley (1743–1805) laid out his exposition of natural theology and developed to its fullest the analogy of the watchmaker, where he contrasts the inference one can produce when encountering a stone or a watch; in the second case, he wrote that the inference of design is straightforward because an inspection of the watch would have us perceive that "its several parts are framed and put together for a purpose, *e.g.* that they are so formed and adjusted as to produce motion, and that motion so regulated as to point out the hour of the day." But, most important, he wrote that this first inference in turn points to a second inference, "that the watch must have had a maker; that there must have existed, at some time, and at some place or other, an artificer or artificers who formed it for the purpose which we find it actually to answer; who comprehended its construction, and designed its use."[101]

By the beginning of the nineteenth century, "the comparison was commonplace, but Paley's late and widely-read treatise gave it its most sustained development."[102] Paley's book was mandatory reading on both sides of the Atlantic Ocean, at Cambridge and Harvard alike. In his autobiography, Darwin wrote, "the logic of his *Natural Theology* gave me as much delight as did Euclid." By the end of his life, however, he seemed to have changed his mind, and concluded: "There seems to be no more design in the variability of organic beings and in the action of natural selection, than in the course which the wind blows. Everything in nature is the result of fixed laws."[103]

Darwin's ideas were indeed so revolutionary that he, like most of his contemporaries, was struggling with their consequences. Later on, it became obvious to many that they amounted to a major discontinuity in human thinking, the second discontinuity according to Mazlich,[104] what Freud had dubbed "the second deep narcissistic wound." Most crucially, Darwin's "dangerous ideas" had the potential to profoundly question the established order of thinking, challenging the respective place of Man, Nature, and God: it was a metaphysical revolution. Darwin, of course, pondered about that, on his own, but also at the explicit request of some of his friends, critics, and adversaries.

Like so many a man of his time, Darwin was religious by default: he did not question his (relative) orthodoxy until quite late in his life. Instead, he assumed for a long time that his theories were not at odds with the revealed theology he was schooled in. He actually believed in some kind of a design argument long after he first thought about the idea of "natural selection." In an intermediary phase in his beliefs, even the "fixed laws" of evolution were not at odds with natural theology, but rather displayed some sort of belief in "general providence." This was, for instance, the idea of his American friend Asa Gray, who tried to defend him from the worst accusations raised against him rather than against his theory: "we should not like to stigmatize as atheistically disposed a person who regards certain things and events as being what they are through designed laws (whatever that expression means), but as not themselves specially ordained, or who, in another connection, believes in general, but not in particular Providence."[105]

Dov Ospovat has argued successfully that such a "belief in general providence" in the form of "designed laws" was indeed Darwin's position at the time of the publication of *On the Origins of Species* (1859) and for some years after. In a note commented by Ospovat, Darwin wrote in 1846: "all allusion to superintending providence unnecessary . . . *rather expressly mention the design displayed in retaining useless organs for further modifications as proof of supervisal.*"[106] The notion that there must be an intelligent designer because of the remanence of some "useless organs" is most notably a part of today's intelligent design theory that applies to junk DNA. The notion that the "purpose" of something (i.e., organ, cell, gene, etc.) could be determined by the possibility of future need was thus not at odds with Darwinism, even if it was a problem for the "rigidly functional viewpoint of most natural theologians."[107] In 1860, Darwin could still believe that "everything [is] resulting from designed laws, with the details, whether good or bad, left to the working out of what we may call chance."[108] By 1871 and the first edition of *The Descent of Man*, however, he had changed his mind and wrote that he was "not able to annul the influence of my former belief, then almost universal, that each species had been purposely created; and this led to my tacit assumption that every detail of structure, excepting rudiments, was of some special, though unrecognized, service."

During the course of his life, Darwin thus went from "unconsidered orthodoxy" to deism, theism, and eventually agnosticism, as he wrote to his friend T. H. Farrer near the end of his life: "[I]f we consider the whole universe, the mind refuses to look at it as the outcome of chance—that is, without design or purpose. The whole question seems to me insoluble."[109] The first move in changing belief was to consider God as having vacated the premises, what Asa Gray called Darwin's adoption of Lord Bacon's "confession of Faith":

> Notwithstanding God hath rested and ceased from creating, yet, nevertheless, he doth accomplish and fulfill his divine will in all things, great and small, singular and general, as fully and exactly by providence as he could by miracle and new creation, though his working be not immediate and direct, but by compass; not violating Nature, which is his own law upon the creature.[110]

Current ID theorists take issue even with this quite standard way to reconcile science with faith, and find in it the origins of an unavoidable drift toward atheism. William Dembski, for instance, wrote "Darwin's theory is a logical result of what happens when one takes natural theology and uses the wrong metaphor, such as the watch metaphor. It takes us away from even deism, and so with Darwin we end up with agnosticism."[111] Moreover, some went one step further and saw in this position the mark of Gnosticism: "this separation of God and the world is one aspect of Gnosticism," wrote C. G. Hunter, and added, "Gnostic ideas predated and influenced the development of evolution, and the wide acceptance of evolution, in turn, strengthened modern Gnosticism."[112] This, of course, was the risk natural theology was facing from its origins: the idea according to which the existence of God could be proved by virtue of his creation, faced with the problem of the existence of evil in Nature, could dangerously lead to either Gnosticism or pantheism, two of the most feared heresies of old. In the first case, the Gnostic theodicy would ascribe the existence of evil to a demiurge, an evil God responsible for the Creation, but not the real true God. In the second case, God would be all in his Creation. This argument was in fact so obvious that it was already stated in the first recorded use of the expression "intelligent design" in 1850—nine years before the publication of *On the Origins of Species*. In *The Theory of Human*

Progression and Natural Probability of Justice, Patrick Edward Dove reconsidered (anonymously) Paley's argument of the two inferences and concluded that the second is "illegitimate until it has been determined what a designer is, and what the term design is really employed to signify." He added:

> If we assume the designer because there is design, we have assumed only a truism; but we have forgotten to establish the most essential proposition, namely, that the adaptation of means to an end *is* design. Every *merely* physical argument to prove the intelligence of the primordial force will split on this rock; and it is absolutely necessary, therefore, for man to progress beyond matter science before natural theology can be other than pantheism. Pantheism is the theology of physical science; and if there were no other science beyond physical science, pantheism would be the last final form of scientific credence.[113]

How clear and rigorous can you get? All but one development was thus obvious from the start, and this development was truly unfathomable for Dove's (and Darwin's) contemporaries: the total disenchantment of the world and the death of God. Only a few years later, it would require such a deranged and genial mind as Nietzsche's to first intuit it. Today all this is again so trivial that none seems to pause when some actually recycle the design argument to man's profit. Consider Elliot Sober's prediction, for instance, "that human beings will eventually build organisms from nonliving materials," an achievement that would not, according to him, "close down the question of whether the organisms we observe were created by intelligent design or by mindless natural processes." Instead, Sober further predicts that "it will give that question a practical meaning," and concludes with an analogy:

> When the Spanish conquistadors arrived in the New World, several indigenous peoples thought these intruders were gods, so powerful was the technology that the intruders possessed. Alas, the locals were mistaken; they did not realize that these beings with guns and horses were merely *human* beings. The organismic design argument for the existence of God embodies the same mistake. Human beings in the future will be the conquistadors, and Paley will be our Montezuma.[114]

Yes, indeed, analogy has a thick skin. The once measure of the Creation of the Intelligent Designer is now applying for the job . . . and the design argument lingers on, merely changing subject . . . as it has changed name and explicative power on several occasions as times passed over the philosophical and the biological knowledge.

Chapter 2, Mostly Head

From Garbage to Junk DNA, or Life as a Software Problem

> We are embedded in our trash—there is no easy way to leap beyond it and build a utopia without garbage, to address the contradiction between the world's limited resources and our seemingly unlimited ability to manufacture trash. Its production is rooted in survival, represented in every culture, and magnified by economic success. To purge the earth of garbage would be to destroy our own reflection.
>
> —JOHN KNECHTEL, *Trash*

Garbage came before junk, even if garbage is more than often the fate of junk: discarded, refused, banished, thrown away. And yet, garbage came before junk when came the time to qualify this uncanny part of DNA, this part which was resisting the smooth efficiency of code turning into program.

Wasn't it this time too, actually, when design took a new meaning, as in the compound expression *designed obsolescence*? From the heyday of modernization, when Taylorism and Fordism became the new *mot d'ordre*, came this lingering concern, and *how long will it last* became *how long should it last?*[1] Ah, control . . . It is not how much you sell, but rather how often you sell. Quality, quantity, frequency—all questions of planning. Inventory control, cash flow, ins and outs, a smooth production process. And this recurring dream, one customer waiting at the factory output, ready to buy this one product that was planned for his or her timely transaction. Go back home and consume and come back in due time! Captive consumers, chained to the production cycle, a closed loop. And they come back for more.

Organic, of course, is of no worry; it is the model. Produced to be digested, consumed and destroyed, excreted as an accursed share, forgotten and bought again, tomorrow if not the day after. Shit, of course, is our first template for garbage: "garbage is organic. It's formless and it stinks"[2] (and Freud added: this is what money is to us, shit). Organic is a wonderful product whose consumption requires its destruction.

This very dream became: make of any product an organic counterpart. And especially commodities, so-called durable goods: garbage in the making.

This is why "natural" materials are to be preferred; they are biodegradable, or else. Ashes to ashes, dust to dust. Rock to sand, metal to rust.

Then came plastic, the real ordeal. Plastic . . . the acme of modernity, the *summum* of style, the maker of fortunes. Durable, moldable, shiny plastic. What would we do without plastics? Plastic is so modern. Owning plastic is like having a little industry in your hand. It is like belonging here, with the civilized. Plastic is ordered, shaped, informed. Even money became clean with plastics. This little plastic rectangle, which means distinction and income, being a member of the cool clubs of the leisure society. *There are some things money can't buy. For everything else, there is plastic.*

The real problem is a saturated market. An inferior-quality plastic: this is the genial trick, the solution. Planned obsolescence, a nice way to say designed garbage. There is some shit money can buy, for everything else there is plastic garbage. Be a good capitalist, train your market to remain unsaturated (this and niche marketing would become the two major commandments of the new gospel). The irony is that when capitalists finally realized that, they got help from one, then two oil crises (and in between Supertramp timely sang on a dumpster background, *Crisis? What Crisis?*). Priceless.

Then came Fashion for everyone, a democratic revolution that turned garbage into brands of all sorts. Trendy crap for all! But I am getting way ahead of myself here. Let us come back to safer grounds. Garbage is first, offal: remains, carrion, carcass, better yet, entrails. The *Oxford English Dictionary* (2d ed., 1989) adds: "rarely the entrails of a man." Rarely, but sometimes. Tripe, guts, where shit comes from. Garbage is what garbage does.

Garbage qua trash became artsy during the twentieth century, and intends to last well into this one. Lately, two paparazzi turned socioartists (as in sociopaths) systematically exposed the guts of the trash cans of the Hollywood stars: fifteen years of hard and dirty work for twenty photographs.[3] The most surprising is how neat this starry garbage is, almost clean, devoid of anything organic. As if a star could not have any dealings with anything organic. For trash is not garbage, nor is it exactly junk:

> Trash, like junk, is often clean, a matter of well-made paper, plastic, and metal. Like junk, trash includes the malfunctioning, failed, burnt-out, and obsolete. If there's a difference, junk tends to be unwanted yet usable, while trash is used up, spent, exhausted, or obsolete. That's why there can be junk shops and junk sales but not trash shops and trash sales. Something can be unused, or even be unusable, without being trash, for example, a cloud. But the cloud never was usable. Trash once was. Trash always has a past, a use that is spent. Like an old person.[4]

The line between junk and trash is fine and blurry. There is more hope to be put in junk than in trash—this little bit of hope that distinguishes what might from what might not be usable in the future. A subjective feeling that might also have some roots in this past Allen alludes to. There is some affect in junk, a bit less in trash, and a lot more in garbage (albeit of the raw kind, repugnance and disgust). Nostalgia for the future is the lot of junk, a kind of "the strange experience of feeling sympathy for rubbish."[5] For trash, however, the only feeling one can have has to be this sadness, "a melancholy that forces us to notice abandoned things and feel for their plight."[6]

There is yet another word in the semantic domain of garbage that is worth pondering about: waste. Like garbage, it refers to an organic and complementary figure of shit: earth, soil. Waste, indeed, originally refers to the land: an inhospitable place, unsuitable for human habitat. But, as John Scanlan noticed, "as the Middle English lexicon expanded to replace this older sense of the term with equivalents like 'wilderness' and 'desert', new uses of waste emerged that began to indicate moral censure."[7] This pejorative sense became associated with uselessness and improper use, neglect and idleness.

Garbage, and its many related signifiers, trash, rubbish, waste, and junk, thus appear as a crucial modern trope. Scanlan even suggests that "garbage provides a shadow history of modern life where the conditions for its production and the means by which it is rendered invisible cast it as an unwelcome double of the person."[8] Here I argue further that late-modern science inscribed "this unwelcome double of the person" at the very core of the person herself, on her most intimate fiber, DNA. It might not have been totally random, then, when first confronted with the realization that most of DNA could not possibly code for the synthesis of proteins, to conclude that it must be *garbage DNA.* Garbage was indeed the first word that came to the mind of the researchers who had such a realization.

cII **Incipit Junk**

Susumu Ohno (1928–2000), a Korean-American scientist of Japanese descent, is usually credited with coining the expression "junk DNA." Among many achievements, he is famous for intuiting the part that gene duplication plays in evolution. In the preface of his classic monograph titled *Evolution by Gene Duplication,* he wrote:

> Had evolution been entirely dependent upon natural selection, from a bacterium only numerous forms of bacteria would have emerged. The creation of metazoans, vertebrates, and finally mammals from unicellular organisms would

> have been quite impossible, for such big leaps in evolution required the creation of new gene loci with previously nonexistent function. Only the cistron that became redundant was able to escape from the relentless pressure of natural selection. By escaping, it accumulated formerly *forbidden* mutations to emerge as a new gene locus.[9]

According to his biographer for the U.S. National Academy of Science Biographical Memoirs Series, Ohno thus "recognized that this [duplicated] DNA could serve as a powerful means by which new genes or new functions of old genes could be created. This concept had been expressed earlier by Haldane, but the explosion in modern biology and molecular genetics made it possible to assess for the first time the important role that gene duplication played in evolution."[10] Ohno considered that at certain key moments of evolution—such as the transition from invertebrates to vertebrates, for instance—entire genomes got duplicated (it is then technically called *polyploidy*). His intuition, however, remained controversial for a while: "At first, geneticists were either enthusiastic or appalled, but by the late 1980s, they had lost interest in the idea because they simply lacked enough data to support or refute Ohno's theory."[11] Nowadays, thanks to a new wealth of data from mapped genomes, Ohno's proposition has been put to the test, and some claim successfully.[12]

By the time he first proposed the notion of gene duplication, Ohno, like many other researchers working on genome size, had already realized that most of DNA could not possibly code for protein synthesis. He reached this conclusion with the help of quite artisanal methods, weighting cutout pictures of chromosomes spreads in order to demonstrate successive doublings of the amount of chromosomal material along the phylogenetic tree.[13]

But Ohno did not call this "extra DNA" "junk DNA" at first. He called it "garbage DNA" instead. When explaining the process of gene duplication, he focused on so-called *forbidden mutations*, these mutations that result in the loss of the function assigned to a single structural gene locus in a haploid genome.[14] These mutations are forbidden because, in altering the DNA sequence for a particular gene, they condemn their bearers to sure death. Selection makes sure that any offspring carrying this detrimental mutation would eventually disappear. In a polyploid genome, however, the situation is very different, since "policing by natural selection becomes very ineffective when multiple copies of the gene are present."[15] In this case, then, forbidden mutations are tolerated as long as they affect duplicates: "in a relatively short time," wrote Ohno, these "duplicates would join the ranks of 'garbage DNA', and finally one functional gene remains in the genome."[16] He made this point clearer through an analogy in a later publication: "The earth is strewn with fossil remains of extinct species; is it any wonder that our genome too is filled with the remains of extinct genes?"[17]

Ohno did not expand on what he meant by "garbage DNA" here, and one can only infer that he uses this term in his vernacular meaning of the period, in the somewhat pejorative sense referring to both lack of function and further utility: useless and refuse-d. In 1972, however, he made his analogy clearer and first linked "garbage" and "junk": "at least 90% of our genomic DNA," he wrote, "is 'junk' or 'garbage' of various sorts."[18] Note the accuracy of the estimate, *at least 90 percent.* Actually, in another paper published that same year, the estimate was even more precise: "taking into consideration the fact that deleterious mutations can be dominant or recessive, the total gene loci of man has been estimated to be about $3x10^4$," he wrote, and added: "even if an allowance is made for the existence in multiplicates of certain genes, it is still concluded that, at the most, only 6% of our DNA base sequences is utilized as genes."[19] Unlike many, Ohno would not have been surprised by the "low number of genes" revealed by the completion of the Human Genome Project. His estimated 94 percent noncoding DNA was very close to the mark too!

Recent commentators have argued that "'Garbage DNA' proved to be an unsuccessful meme, but its essence remains in the wildly popular term coined by Ohno two years later—'junk DNA.'"[20] But they fail to see that moving from "garbage" to "junk" was a profoundly logical move in Ohno's thinking, responding to very definite imperatives. First, the overall logic of his argument was economic: his estimates dealt with what life could "afford" in terms of deleterious mutations (i.e., "forbidden" mutations). Second, in this logic, "junk" appears to be only one "sort" of "garbage." It is pretty clear, then, that "garbage" and "junk" were not synonymous in his mind.

"Garbage" is the whole encompassing term that refers to all kinds of mutated duplicated gene loci, where forbidden and/or tolerable mutation is the core process of evolution, the differentiation of life. "Junk," on the other hand, refers to some potential positivity for some of this "garbage." The following sections of his paper were very clear about these points, and alluded to at least two different kinds of such positivity: buffering against mutations and functional divergence.

Buffering. Some noncoding sequences ("untranscriptable and/or untranslatable"), he realized, "appear to be useful in a negative way." He labeled this "the importance of doing nothing." By this he meant that these sequences, in spacing genes far apart one from another, buffer them from a deleterious mutation and thus "confine them to a single locus, instead of allowing it to spread to other genes."[21]

Divergence. In this second kind of positivity, the duplicates constitute a kind of reservoir for functional divergence: "the creation of new gene with hitherto nonexistent function is possible only if a gene becomes sheltered from relentless pressure of natural selection . . . Redundant copies of genes thus produced are now free to accumulate formerly forbidden mutations and thereby to acquire new functions."[22]

These intuitions are still being investigated and debated today. I do not intend to cover all these contributions here. Instead, I want to stress two main points: (1) "junk DNA" appears historically as a specification of some kind of "garbage DNA," and (2) the question of its potential function was left open when the expression was first coined.

For the first argument, junk DNA appears as a special class of "garbage." As such, it confirms the implicit meaning embedded in this word: "garbage in all its forms can be said to represent nature (including human existence) as an endless process of generation and decay."[23] What differentiates junk from garbage, again, is the potentiality of further usability, or, in the biological vocabulary, the emergence of a new function, called functional divergence. Garbage, in this perspective, is what is definitely decayed; a remainder, an accursed share with no potentiality for reconnection with the greater whole, since, as Scanlan aptly puts it, "garbage arrives at its fate because it either suffers or effects some disconnection—it is the separation of a part from some greater whole."[24]

For the second argument, one must notice that if the potential function of junk DNA is what differentiates it from "garbage DNA," Ohno did not conclude necessarily that *all* of what is considered today as junk DNA must have a function. If, as Gregory puts it, "garbage DNA" eventually was "an unsuccessful meme," it is because it totally disappeared from the biological lexicon, and its absence thus created some ambiguity:

> From the very beginning, the concept of "junk DNA" has implied non-functionality with regards to protein-coding, but left open the question of sequence-independent impacts (perhaps even functions) at the cellular level. "Junk DNA" may *now* be taken to imply *total* non-function and is rightly considered problematic for that reason, but no such tacit assumption was present in the term when it was coined . . . Its current usage also implies a lack of function which is accurate by definition for pseudogenes in regard to protein-coding, but which does not hold for all non-coding elements. The term has deviated from or outgrown its original use, and its continued invocation is non-neutral in its expression—and generation—of conceptual biases.[25]

In this illuminating commentary, Gregory introduces the word *pseudogene.* This is the technical word used now to refer to what Ohno originally called "garbage DNA." Pseudogenes are "fossil" genes, vestigial sequences, which once had a function (i.e., encoded in DNA some amino-acid sequence specifying a certain protein). The fossil analogy, once made by Ohno, is still frequent in today's literature.[26] The word *pseudogene* was coined in 1977,[27] the same year that the "introns," this "minirevolution in molecular genetics," were discovered. Like almost anything connected to DNA and molecular biology, they soon got translated into the reigning cybernetic metaphor. As programs go, pseudogenes are "backups," as in this metaphoric rendering by William Calvin:

> "So duplicated genes, however they might arise, would be handy for evolving a new improved version of a living thing," Rosalie said. "And there are sure a lot of near-duplicates of genes in the cell nucleus, strings of DNA which are almost like the ones that make the proteins, but not exactly the same."
>
> "Of course, if the changes are made randomly, with no intelligent programmer masterminding the modifications, you'd expect that a lot of the DNA sequences would be nonsense, simply nonfunctional," I added. "Nothing but junk."
>
> "So that's what is called junk DNA?," asked Abby. "Or is it garbage DNA, I forget?"
>
> "Someone once called it garbage DNA, but the more appropriate name is junk. After all, garbage can be thrown away. Junk is the stuff that you never manage to get rid of," Rosalie said, laughing.[28]

Rosalie omits to mention why we never manage to get rid of our junk. Some, for sure, feel an attachment to it, testimony to a moment in their life, a feeling associated with an event from the past. Some others collect it until there is enough of it to attract some attention and eventually sell it in a garage sale. Yet others always believe that they will eventually give a second life to it, mend it, fix it, recycle it to new use. This last option is not devoid of interest when one considers the fate of some pseudogenes. This was, after all, one of Susumu Ohno's original ideas; being relatively free from selective pressure as "backups," gene duplicates can amount to a reservoir for change. But there is yet one more analogy in this exposé, which was promised to a bright future: that of a "programmer." Here was, in fact, the master trope of the molecular biology revolution of the 1960s and 1970s.

Under this renewed trope, even pseudogenes could find a regulatory function, blurring once again the boundary between garbage and junk.

A **A Thousand Loops**

The world became a loopy machine, down to its molecular level. One was used to the straight line, and suddenly—or so it seemed, at first—the line looped. Maybe there was a loophole in the linear text of the Cartesian subject, and the line fell into it and the subject thought he could escape from it, through the loophole. Anyway, suddenly the world became loopy, and so did the subject.[29] The line could not hold him anymore. The idea was here since the Baroque age, at least, but it came back with a vengeance: the loop became the only possible fate of the fold.

Some claim that it was the doing of cybernetics, this revolution of the mechanistic worldview that occurred during World War II. Cybernetics, the word, was not new, by any means. Plato first, in the old age, and André-Marie Ampère[30] (1775–1836) second, at the interface of classical and the modern ages, had already used it. To them, it meant the governance of men, the steering of people. This world, of which we were once the measure, became a loopy machine inasmuch as we, in return, needed to be steered, cared for, disciplined, and punished (let us call this the imperative of mechanical subjectivation). Man, it was already well established, was an automatic machine: "The human body," wrote Julien Offray de La Mettrie in 1748, "is a machine which winds its own springs."[31] What was not that clear, however, was precisely how this spring got wound up. Here came the loop.

Here is a speculation. This machine, whose spring it can wind on its own, must be a clock, or contain some sort of a clock. Life's second wind is this kind you can give to your watch (when you give it a wind). This spring evokes a (mortal) coil, and the watch reminds us of Paley. Life is this kind of mechanism that would go on forever as long as it is wound. The loop might be the revolution of the hands of the watch, or it might be this time between the first and the last wind. But it is not the way the loop got in, I think, even if I agree with Lewis Mumford on this: "the clock, not the steam engine, is the key-machine of the modern industrial age."[32]

The loop, it is said, came with the steam engine and its regulator, Watt's governor.[33] Thus automation was born practically first, as *déphasage,* as Gilbert Simondon would have put it. It was, of course, a question of a well-adjusted behavior, regular performance *(like clockwork, this bourgeois ideal).* Under a new name, the loop became one of the key concepts of a formidable synthesis, uniting animal and machine under the auspices of this great mechanism: *feedback.*

Feedback is another name of the loop, the technical name under which went this other key concept of cybernetics. Cybernetics, the science of communication and control, rests on these two pillars: a theory of communication (information and code) coupled to a theory of control (feedback). Control is of major importance; it is the insurance of performance, the process of maintaining equilibrium or aiming toward something. Thus control and regulation go hand in hand: "when we desire a motion to follow a given pattern the difference between this pattern and the actually performed motion is used as a new input to cause the part regulated to move in such a way as to bring its motion closer to that given pattern."[34] This, however, is only the definition of *negative* feedback.

Prior to this definition in the eponym monograph that manifested the birth of cybernetics, the concept of feedback had already been at the center of another paper, which, in turn, has been called the "birth certificate of cybernetics."[35] This 1943 paper by Arturo Rosenblueth, Norbert Wiener, and Julian Bigelow, titled "Behavior, Purpose and Teleology," was no less than the original manifesto for a new conception of teleology based on the notion of negative feedback. It plainly equated both notions, in fact: teleology became "synonymous with feedback controlled purpose." This, of course, was very problematic.

Teleology usually means "the study of the evidences of design in nature . . . a doctrine explaining natural phenomena by final causes . . . the fact or character attributed to nature or natural processes of being directed toward an end or shaped by a purpose."[36] Teleology thus usually requires a type of causality in which the effect is explained by an end (in Greek, *telos*) to be realized, "that for the sake of which a thing is done." This is Aristotle's final cause,[37] the most crucial cause, according to him. Teleology differs essentially from efficient causality, "the primary source of the change or rest," in which an effect is dependent on prior events.

Teleology, as we have seen, is at the core of the design argument, the very foundation of natural theology (the teleological argument in Aquinas's fifth way). Modern science, on the other hand, thrived on the elimination of the final causes, "perhaps the least common denominator in the various conceptions of the mechanization of the universe."[38] Wiener and his colleagues were still modern scientists: they did not seek to reintroduce final causes in the explanation; instead, they wanted to reintroduce "purpose" without the burden of sequential causality: "Since we consider purposefulness a concept necessary for the understanding of certain modes of behavior," they wrote, "we suggest that a teleological study is useful if it avoids problems of causality and concerns itself merely with an investigation of purpose."[39] In order to do so, they used another Aristotelian idea on causality, that of *circular causality*, where "some things cause each other reciprocally, e.g. hard work causes fitness and vice versa." Aristotle, however, had insisted that in such case, the reciprocity was asymmetric: "but again not in the same way, but the one as end, the other as the origin of change."[40] They obviously did not bother with this proviso.

There is thus a logical problem at the core of the first loop. Feedback still implies "an origin and an end of change"; it is the output that is fed back. In overlooking this problem, their argument ran the risk of tautology: "since we are not permitted, so to speak, to look inside the object, the only way in which we can tell whether or not it has been modified is to observe a modified aspect of its external behavior. But behavior is defined in terms of output and input. Thus behavior is defined in terms of behavior. What is wrong?"[41]

This tautological argument had also another consequence of major importance for the future of the loop: it lumped together equilibrium and goal-seeking mechanisms—"signifying homeostasis as negative feedback and then resignifying such servomechanisms as organismic homeostasis amounted to circularity."[42] So the logic governing the introduction of the first loop was itself loopy. If such was the case upstream, it would soon follow that it would be the case downstream too. From one loop to the infinity of loops, it became a loopy world, without origin or end: a world best conceptualized through the eternal return (of the loop).

In biology, the loop had a great future, as great as teleology had a great past. But for this future to exceed this past, one first translation was needed: from teleo*logy* to teleo*nomy*. This move in the suffix, from *logos* to *nomos*, is highly significant. In one sense, teleonomy can be taken as merely synonymous with teleology. In this case, the suffix *–onomy* is plainly this suffix which names certain disciplines of study, such as astronomy, or economy, when other disciplines are named with the help of the synonymous suffix *–ology* (such as biology or sociology): both suffixes mean "the study of." However, when one looks in detail at which disciplines are named with which suffix, it is obvious, that disciplines in *–onomy* refer more often than disciplines in *–ology* to the study of a phenomenon governed by some laws. This is especially obvious in the cases where there exist two "disciplines" with the same prefix: think about the differences between "astrology" and "astronomy." From this case one can easily infer that *–onomy* refers to the "scientific study" and *–ology* to the mere study of a given phenomenon. Or, in other words, *-onomy* is one level higher than *–ology* in the scale of intellectual respectability. Indeed, it seems that the question of "respectability" might have been at the core of the lexical move in biology; or such seems to be the recollection of the man who is usually credited with coining the term:

> The apparent inseparability of purpose and consciousness was previously responsible for a major embarrassment and impediment to the biologist that Haldane put in a characteristically pithy way, "Teleology is like a mistress to the biologist; he dare not be seen with her in public but cannot live without her." It was my intention in 1957 to help get Haldane's mistress out of the closet by describing her merits as teleonomic rather than teleological. Whether or not that did help (Monod & Davis found it useful!), the commonplace nature of programmable machines at midcentury gave teleology (as teleonomy) complete respectability in the society of biological ideas. The genome was the program of a Turing machine and Darwin's Demon was the programmer.[43]

François Jacob, some years later, was even more straightforward in the same pithy way when he claimed that "the concept of program has made an honest woman of teleology."[44] And some more years later, Richard Dawkins reascribed this "respectability" to Darwin himself, when he defined teleonomy as "the science of adaptation": "In effect, teleonomy is teleology made respectable by Darwin."[45] The notion of "programmed purposiveness" actually reconciles and unites these two claims. It is, in effect, the major tenet of this molecular synthesis that went from Wiener and his colleagues, through Pittendrigh, Mayr, Jacob, and Monod, and eventually to Dawkins and Crick and his colleagues.

Ernst Mayr might have first enunciated the new notion of "programmed purposiveness." In a November 1961 paper entitled "Cause and Effect in Biology: Kinds of Causes, Predictability, and Teleology Are Viewed by a Practicing Biologist," he adopted Pittendrigh's term, but offered a key precision: "it would seem useful to rigidly restrict the term *teleonomic* to systems operating on the basis of a program of coded information."[46] This was a particularly shrewd move, because it entailed considering that life has a purpose resulting from the functioning of a program that appeared *prior* to the purpose: in other words, purpose is here a *descriptive* rather than a prescriptive notion. This is exactly in tune with the cybernetic redefinition of purpose[47] as programmed purposiveness, and as such it had a crucial consequence for the ontology founding the research program of molecular biology: it enshrined the metaphor of DNA qua program. As Pittendrigh had put it early on, "the primary purpose of biological organization is self-perpetuation by self-copying, and as such is the handiwork of Darwin's Demon . . . the computer revolution for which he was so largely responsible had a major impact on the climate of biological thought at mid-century."[48] Darwin's demon *is* this programmer, and for creating it, Darwin is legitimately credited with the renewed respectability of teleology in biology, under the new name of teleonomy.

One consequence is clear according to this retrospective reading: from Darwin to the neo-Darwinian synthesis revisited by molecular biologists post-World War II, finality became inherent to the representation of life as *teleonomy,* a new guise for the Darwinian adaptation. In the new framework of molecular biology, natural selection would still be the core mechanism of evolution, but redefined as a conservative process prone to errors (i.e., mutations). Everything would eventually rest on the self-perpetuation of DNA, its *invariance.* Prior to Dawkins's notion of selfish gene revamped as selfish DNA by Crick and his colleagues, it would be nowhere as clear as in the work of the "Pasteur connection," which added one more degree to the loop.

In the early 1960s, even before the genetic code got completed, a team working at the Pasteur Institute in Paris completed the cybernetic paradigm in biology by working on its second pillar, a theory of genetic regulation. François Jacob (b. 1920), Jacques Monod (1910–76), and André Lwoff (1902–94) shared the 1965 Nobel Prize in physiology or medicine for their "discoveries concerning the genetic regulation of enzyme and virus synthesis." These discoveries were of two kinds: the role of messenger RNA in the synthesis of proteins and the operon model of genetic regulation. Both aspects of the work of the Pasteur connection have been very well covered in recent works in the history of biology.[49] Their contribution can indeed be considered as a founding moment in the life of molecular biology because it buttressed the "genetic program" metaphor. In fact, their work on the operon model provided an answer to the most crucial questions left open by previous discourses on "programmed purposiveness" (i.e., teleonomy) in biology: "Are genes to be understood as the subject or the object of the genetic program? Are they the controllers or the controlled, the regulators or the regulated, the switches or the entities to be turned on and off, the activators or the activated?"[50] They answered, to their fame, *both, my captain!* In the first case, they decided that genes ought to be called "regulatory genes," and in the second case, "structural genes." With the invention of regulatory genes, they were thus able to "redefine both feedback and regulation (and even epigenesis) to refer to genetically controlled processes."[51]

Moreover, this clever solution permitted keeping the whole field of molecular biology inside the set of principles that had governed it since the discovery of the structure of DNA, some ten years earlier. This solution confirmed the "Central Dogma" and still gave a crucial role to proteins, and especially to a special class of proteins, namely, *enzymes*. The most elementary feedback possible would have been one where the end product of the activity of a structural gene, the protein it encoded in DNA, would be retroactively regulating the activity of the given gene. Jacob and Monod instead proved that only certain proteins act accordingly, not for their own synthesis, but rather as catalysts for the synthesis of other proteins. This is why they called their work "cybernétique enzymatique." Thus they introduced a loop inside the loop, leaving open the possibility for other loops to come: "it was of course understood that, in order to do their job and turn the structural genes on and off, regulatory genes themselves needed to be activated, but this fact seemed to present no impedance whatsoever to the new construction."[52]

From then on, François Jacob could rightly announce, "Today biology is concerned with the algorithms of the living world," inasmuch as the notion of a "genetic program" had become the central tenet of the new paradigm of molecular biology. This, however, was not devoid of a constitutive ambiguity: if there is still a debate about the metaphoric nature of the use of the notion of "code" to characterize DNA, there is not much doubt that the use of the notion of program is here essentially metaphoric, and thus comes with the uncertainty, the productivity, but also the downfalls of such a mode of knowledge, as Evelyn Fox Keller has so well argued: "here, too . . . productivity also had its downside."[53]

We know now where this downside lies: in the erroneous one-way information transfer of the Central Dogma, the excessive primacy of proteins for regulation purposes, and the dead end of collapsing purposiveness with the will to replicate invariantly, best expressed in the selfish DNA fiasco of the 1980s. If the primacy of proteins to characterize the chemical mechanisms of life itself had indeed a downside, it was of course most obvious in dealing with this accursed share that junk DNA represents for this model. The difference, however, might not have been so irreconcilable as it seemed at first sight, and that too offered a new way to sort junk out of garbage.

B **Regulation, without a Program**

This is where the work of Roy J. Britten, the marginal orthodox whom we already met in chapter 1, comes into play. Unlike Ohno, Britten did not confuse junk with garbage; moreover, he is often quoted as having first enunciated the central axiom of junk (often wrongly attributed to Sydney Brenner, who seemed to have rediscovered it quite lately): "Garbage you throw away, junk you keep just in case . . ." Even prior to Ohno's coining of the "junk DNA" expression, Britten was busy working on "repetitive DNA," the only legitimate name for most of junk DNA, according to him. In two seminal papers, in 1968 and 1969, followed by multiple others as the years would go by, he laid the foundations of what amounts today to a new paradigm in genetic regulation. This new paradigm, if it were summarized in one slogan, would be: "Down with the genetic program, long live the regulatory networks!" This new paradigm, if it were summarized in one key idea, would be: "Genetic regulation is not carried out by proteins only, but rather by very complex networks where noncoding RNAs (i.e., transcribed but not translated DNA sequences) play the key parts." Now, you guessed it: noncoding RNAs are the transcripts of formerly considered junk DNA.

Roy Britten was born in Washington, D.C., in 1919. His mother worked at the National Research Council and his father was a statistician; he was thus exposed to science early in his life. In 1940, he graduated from the BA program in physics at the University of Virginia. Not long after, he was recruited to work on the Manhattan Project, where, according to him, "I did not do anything useful."[54] He didn't return to graduate school until 1946, when he enrolled in the nuclear physics program at Princeton. He graduated in 1951, but soon decided that he did not want to continue working in this field: "it became apparent to me that if I were going to work in that area, I would be part of an enormous team of people . . . I did not want that."[55] Like many physicists with no taste for "big science" after World War II, Britten thus moved to biology. He took the legendary phage course established in 1945 by Max Delbrück and Salvador Luria at Cold Spring Harbor Laboratory, started his postdoctoral work at the Carnegie Institution of Washington (CIW), and eventually stayed there until 1973.

With his colleagues from the Biophysics group at the Department of Terrestrial Magnetism at CIW, Philipp H. Abelson, Dean B. Cowie, and Ellis. T. Bolton, under the direction of physicist Richard Roberts,[56] Britten participated in the 1950s in "the enzyme period," which preceded the modern era of molecular biology: "They pioneered the use of radioisotopes for the elucidation of metabolic pathways, resulting in a monograph[57] that guided research in biochemistry for the next twenty years and, together with early genetic and physiological studies, helped establish the bacterium E. coli as a model organism for biological research. During this time, most of the metabolic pathways required for the biosynthesis of intermediary metabolites were deciphered and biochemical and genetic methods were developed to identify and characterize the enzymes involved in these pathways."[58] According to Britten, "while the term 'feedback inhibition' was devised by others, its existence was proved by the work of the group during this period."[59] The loop was thus already at the center of their work, at the same time that Jacques Monod and his colleagues were busy working on it.

In the mid-1960s, Britten and the other members of the group started working on DNA hybridization kinetics. Before the advent of genetic engineering, hybridization was the only way to provide global information on the organization of the genome.[60] The key to this technique is the spontaneous reassociation of the separated strands of DNA in experimental conditions (i.e., in vitro solution), observed and studied in the group at that time.[61] In 1964, they observed the reassociation of vertebrate DNA, and they proved that the extent of the reassociation between DNA strands derived from *different* species was a measure of the evolutionary relationship between the given species.[62] Britten's original insights on these studies led him to make the hypothesis of the presence of frequently repeated sequences in the DNA of vertebrates.[63] This was soon confirmed in a series of experiments and papers spanning the years 1965–68, leading to a major 1968 paper in *Science*, coauthored with David E. Kohne, then a postdoc in the group.[64] In the conclusion of this paper, Britten wrote:

> A concept that is repugnant to us is that about half of the DNA of higher organisms is trivial or permanently inert (on an evolutionary scale). Furthermore, at least some of the members of DNA families find expression as RNA. We therefore believe that the organization of DNA into families of related sequences will ultimately be found important to the phenotype. However, at present, we can only speculate on the actual role of the repeated sequences . . . The DNA of each vertebrate that has been examined contains some families with 100,000 members or more. This very large number suggests a structural or regulatory role.[65]

This was his crucial insight, not altogether original, but worthy of notice, since Britten stuck to it, against the general and authoritative consensus of the time.[66] This "act of faith," based on true speculative thinking, then led to the development of a theoretical model, also published in a major paper in *Science* the following year.[67] Britten still considers today this paper as a bet against the odds: "That came from a challenge. Davidson and I made the challenge that led to that 1969 paper."[68] Eric H. Davidson (b. 1937) was then a young assistant professor at Rockefeller University, which he had joined after his postdoc with Alfred Ezra Mirsky, in June 1965. The connection with Mirsky proved to be essential.

Alfred Ezra Mirsky (1900–1974) is rightly considered a pioneer in eukaryotic molecular biology, and especially of proteins studies. His early interest in protein denaturation led him in 1936 to the California Institute of Technology, to work with Linus Pauling on a hypothesis of protein folding and stabilization by hydrogen bonds. In his later collaborative studies with Hans Ris, he provided some of the evidence for DNA as hereditary material by showing that diploid cells have a constant DNA content, whereas haploid cells have half as much DNA (the "C-value"; see chapter 1). In the 1950s, he pursued his interest in proteins and began to investigate their biosynthesis in cells. Nucleoprotein chemistry and developmental biology became major themes in his laboratory at the Rockefeller Institute, and he developed the hypothesis of "variable, environmental control of gene expression," that is, how the same hereditary material in all cells can give rise to the differentiated organism and remain involved through the life span in responses to environmental change.[69] By the end of the 1960s, the concluding works of his laboratory were concerned with problems of embryological development, observations on the modification of histones by acetylation and methylation, and on the effects of such substitutions on gene expression, and many other aspects of the contents and activities of cell nuclei. Mirsky's last papers, published in the early 1970s, focused on the role of the histones in the structure of chromatin and in its replication and transcription.[70]

During his postdoc in Mirsky's laboratory, Eric Davidson began working on cytoplasmic factors regulating gene activity during early development; his article for *Scientific American* (1965) on hormones and gene activity represents another outgrowth from the far-reaching laboratory discussions.[71] According to Britten's recollection of the beginning of their collaboration, "Davidson was working in developmental biology when he realized that the kind of experiments he was doing had a problem with repetitive DNA. So he came down to learn about it."[72] This "problem" was discussed in the first sentences of the introduction of their subsequent paper:

> Cell differentiation is based almost certainly on the regulation of gene activity, so that for each state of differentiation a certain set of genes is active in transcription and other genes are inactive. The establishment of this concept has depended on evidence indicating that the cells of an organism generally contain identical genomes. Direct support for the idea that regulation of gene activity underlies cell differentiation comes from evidence that much of the genome in higher cells is inactive and that different ribonucleic acids (RNA) are synthesized in different cells.[73]

In the first footnote to this introductory paragraph, they referred explicitly to Mirsky's work, and gave him joint credit for having formulated first in

1950 "the variable gene activity theory of cell differentiation" (the aforementioned "concept"). They proceeded then to expose their own model of cell regulation. Their crucial point was to postulate processes *at the level of genomic transcription only*, therefore focusing on RNA agency and thus leading to considering nontranslated RNA (i.e., noncoding RNA) at the core of the model. Apart from this crucial difference, their model looked admittedly similar to Jacob and Monod's operon model. They exposed very careful justifications of its most heterodox aspect, that previously considered junk parts of DNA herein acquired a regulatory function: "the genome of an organism can accommodate new and even useless or dangerous segments of DNA sequence such as might result from saltatory replication . . . Initially these sequences would not be transcribed, and thus would not be subject to adverse selection."[74] This character of the model alluded to a mechanism of the creation of an evolutionary reservoir for change stemming from gene or genome duplication first proposed by Ohno, but crucially adapted it to the mechanism of a sudden amplification of a DNA sequence to generate many copies in a tandem arrangement ("saltatory replication"). Moreover, the model predicted a regulatory function for repetitive DNA, as the origin of "a certain class of RNA molecules . . . confined to the nucleus . . . not precursors of cytoplasmic polysomes."[75] In other words, transcribed but not translated DNA.

This was highly heterodox, since it was already known at that time that most repetitive sequences stem from transposition inside the genome. This aspect of the model thus directly referred, among other lines of evidence, to the work of Barbara McClintock, who had "demonstrated the presence of other control sites adjacent in the genome to the same producer genes as those controlled by the distant regulatory elements."[76] In our interview, Roy Britten confirmed that they were aware of the heterodox nature of McClintock's work in the field at that time: "I don't think Eric and I referred to her properly, although we did refer to her. We flipped a coin as to who was going to go and discuss it with her, where the situation was so we could properly include it in the paper. At that time, geneticists just ignored it, because it was outside the paradigm."[77] Britten has now proved that McClintock's work and his intuition of the importance of transposable elements were very significant: apart form their role for regulation purposes, he published in *Proceedings of the National Academy of Sciences* in 2004 and 2006 two papers showing that transposed sequences can contribute to the apparition of novel genes.[78]

The 1969 paper, with its highly heterodox aspects, was thus the evidence of Britten's (and Davidson's) pioneering work in the study of the repetitive sequences of junk DNA. It was not exactly welcome inside the standard paradigm, characterized by the Central Dogma and the preeminence of proteins, and even less in its supplemented version with Dawkins's selfish gene turned Crick et al. "selfish DNA." Even if Britten and Davidson kept on working and publishing on this line of inquiry,[79] formal recognition would come only much later, when some astonishing results proved that their theoretical model was right on the money.

This set of results led to the discovery of the phenomenon known as "gene silencing," the general term used to describe the epigenetic processes of gene regulation (the "switching off" of a gene by a mechanism other than genetic mutation). We know now that genes may be silenced at either the transcriptional or the posttranscriptional level. Transcriptional gene silencing is the result of modifications (methylation, acetylation, etc.) of the histone proteins, creating an environment of heterochormatin around a gene that makes it inaccessible to transcriptional processes. Posttranscriptional gene silencing is the result of mRNA of a particular gene being destroyed. The destruction of the mRNA prevents translation to synthesize a protein. A common mechanism of posttranscriptional gene silencing is *RNA interference,*[80] a mechanism for RNA-guided regulation of gene expression in which a double-stranded ribonucleic acid (dsRNA) inhibits the expression of a gene with a nucleotide sequence complementary to that of the given gene. The phenomenon of RNA interference, broadly defined, also includes the endogenously induced gene-silencing effects of microRNAs (miRNAs), a kind of noncoding RNA particularly active for cellular differentiation during development of the organism.[81] Most of these results were published during the last ten years of the twentieth century. R. C. Lee and his colleagues first described microRNAs in 1993.[82] The term "*miRNA*" itself, however, was first introduced in a set of three articles in *Science* in October 2001.[83] Andrew Fire and Craig C. Mello first published their work on RNAi in the nematode worm *C. elegans* in 1998,[84] and were eventually awarded the Nobel Prize in physiology or medicine in 2006 for this work. If the value of Britten and Davidson's pioneering model took more than thirty years to be confirmed with experimental results, this recognition was, however, clear and groundbreaking in its effects. Some have even argued that it amounts to the birth of *a new paradigm in developmental biology:*

> Thus it appears that the genome is largely composed of sequences encoding components of RNA regulatory networks that co-evolved with a sophisticated protein infrastructure to interact with RNAs and act on their instructions . . . The irony is that what was dismissed as junk because it was not understood may

well comprise the majority of the information that underpinned the emergence and now directs the development of complex organisms.[85]

Several key elements, however, relativize the scientific revolution implied in the affirmation of a new paradigm. Crucial aspects of the "new paradigm" remained unchanged. Foremost among them is the continuing presence of the loop, that is, the centrality of the cybernetic concept of negative feedback. At this level, molecular biology has moved from an elementary model where the protein product itself is fed back to regulate its synthesis, to the recognition of regulatory genes whose products are involved in the "double negation" of the operon model, to, eventually, a thousand loops, the network of complex regulatory interactions of RNA interference. Figure 6 shows an example of such an end point in the conceptions of the loop.

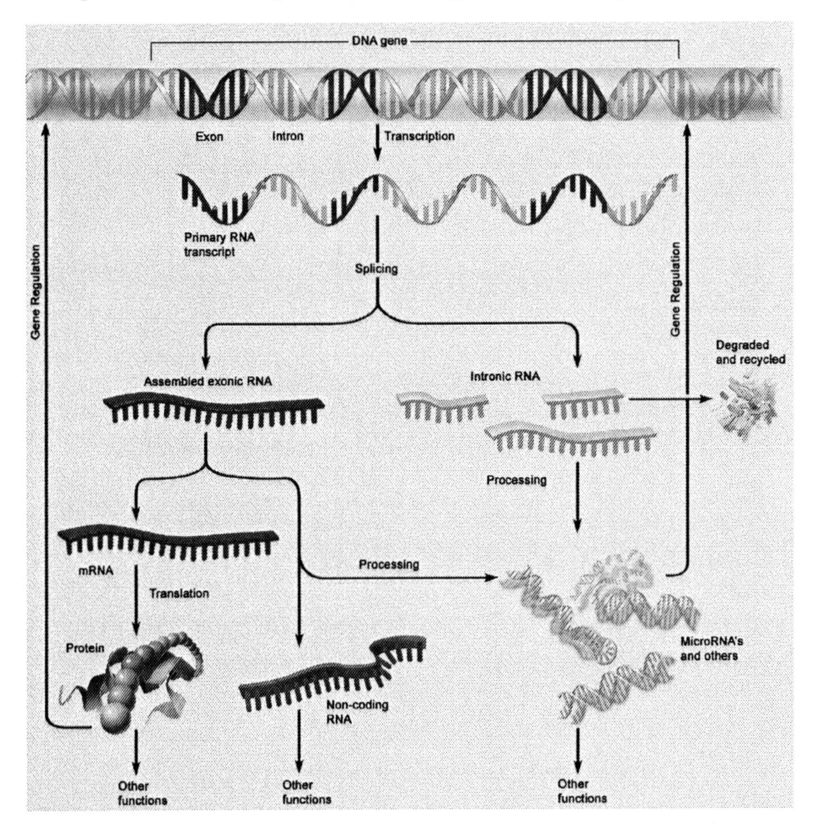

Figure 6. The elementary scheme of networked regulation. From John S. Mattick, "The Hidden Program of Complex Organisms," *Scientific American* 60–67 (October 2004): 65. Copyright 2004 Terese Winslow.

Moreover, the continued use of the metaphor of "genetic programming" still clearly inscribes the "new paradigm" in the tradition of cybernetic molecular biology. DNA is thus still envisioned as a Turing machine or as a set of related Turing machines, in spite of some paradoxical aspects of this conception. Indeed, the importance of the mobile elements (transposons, retrotransposons) inside the genome appears quite at odds with this metaphor, even when assuming the form of a network of "subroutines" rather than a single linear program. In this perspective, Roy Britten answered quite categorically when I asked him which vocabulary was adequate to characterize "junk DNA":

> TB: Would you use the word parasite or parasitic?
> RB: Parasitic? Yes. That word is my preference.
> TB: Program?
> RB: Probably not.
> TB: Purpose?
> RB: It's dangerous. Does it have a purpose? No, I think in terms of evolutionary theory the word purpose is better left out.
> TB: Regulation?
> RB: Of course, it is a subject I worked on for many years, gene regulation.
> TB: Selfish?
> RB: People do use it . . .
> TB: Would you use it? If people do, would you?
> RB: Only when I was trying to suppress what Crick said . . .
> TB: Now, would network be a word that I should add to my list?
> RB: I have been thinking of it lately. Yes, there is no question it should be part of it because there is a network in gene regulation and there is a network for gene interaction.[86]

One should thus note in these answers that the metaphor of the network does not necessarily imply in Britten's mind the use of the program metaphor. It may be considered ironic that the most crucial performativity of the program metaphor in molecular biology might have been in the production of another metaphor that did not require its being cast in the programming framework in the first place. Let a thousand loops bloom. From a different perspective, Britten's answers also told me that the teleological debate was far from over—for better or worse. Could it be soon the last remaining "reason" for junk to remain selfish?

C **Bioinformatics**

"The disembodied Agent then—as near to God as makes no difference—is a spirit, a ghost or angel required by classical mathematics to give meaning to 'endless' counting."[87] The modern computer is this disembodied Agent, ghost, or angel. It is it, or they, when networked, that counts endlessly for us—moreover, that count us, endlessly, since the beginning of their inception. The computer was first the master of the census, this age-old practice of writing the numbers of the enslaved stock, turned mechanistic (writing was first counting slaves and other livestock).[88] The automated census and its punch-card mechanisms (already IBM) marked the beginning of the era of the control society (see chapter 4), no more paradise than hell, a kind of purgatory, in fact. For a century and change it has counted us, classified us, sorted us, named us, and thus helped discipline and punish, "steer" or control us. After a while, we got so used to it, to our *matricules*[89] and other reference numbers, that we could imagine without any trouble that we actually *were* them: I am this address, this finite set of digits (this is the synopsis of the script of how we got databased).

After all, the analogy went both ways; they were first our giant brains, our exocortexes, and thus they soon became our own idea of our brains and/or minds. The mind is a computer and the computer is a mind. A computer is a computer is a computer. This is how we got computed. For a while, this was a metaphor, a mere metaphor, and, as everybody knows by now, "the price of metaphor is eternal vigilance."[90] The price the computer paid, indeed . . . Turned into a machine, the counting Agent got a taste of the infinite at the cover price of losing his or her body, and the duty of perpetual vigilance. Let it be our vigil, for the created eternity, that is, perpetuity: a purgatory indeed. If man is this self-wound spring machine, a perpetual clock until death does him apart, there must be a watcher. And eventually, some argue, a watchmaker, nah . . . a computer will do the job: "they needed a dancer, but a calculator got the job, poor Figaro!" There was indeed some rationality in this choice. Job, it is said, comes from *gob,* "a mouthful, lump,"[91] and in slang also refers to the person herself. The computer, this brainy job.

But then again, the fate of the successful metaphor is to turn into a world. And so it happened this time too, and we, literally, got computed. We are now officially databased to the bone, to the most minute fingerprint, at the molecular level: our DNA print is our new brand. This is no more metaphor, even if out of control. Why bother with control if you can actually make the entity to be controlled? Heredity, the mark of a free man, turned out to be computed at his molecular level, in his cells, in the nucleus of his cells, his DNA (reminder: the slave has no name, i.e., no heredity). And in the last half of the last century of the second millennium, genetics, the science of heredity, turned into an information science, and then into a computerized science: *bioinformatics.* No wonder, then, that the human being himself became *biocapital* in the same process.

Bioinformatics is, quite simply put, the technological engine of this process, "the application (or integration) of computer science to molecular biology."[92] Eugene Thacker described it also as an "ontological practice . . . participating in the reconfiguration of dominant ways of understanding the relation between the living and the non-living, the biological and the technological . . . not just an informatics view of biology; it is also at the same time a biological view of economic value. This twofold character of bioinformatics—at once ontological and political—can be called *the politics of 'life itself.'*"[93] This happens to be the title of a recent book by Nikolas Rose, who described five "pathways" to characterize present-day biopolitics as these "politics of life itself": (1) molecularization, (2) optimization, (3) subjectification, (4) somatic expertise, and (5) economies of vitality.[94] The first two pathways are technologically determined—although, as a good Foucauldian, Rose would deny such an abrupt statement, to refer instead to "hybrid assemblages of knowledge, instruments, persons, systems of judgment, building and spaces, underpinned at the programmatic level by certain presuppositions and assumptions about human beings . . . [and] oriented toward the goal of optimization."[95]

Here I am going to be less complacent with the fashionable discourses of "hybrid assemblage" in poststructuralist social studies, to focus instead on *the* major assemblage, if there were any, that of biologists and computers. After all, Lincoln Stein once defined the field in the most minimal way as "biologists using computers or the other way around."[96] And I intend to be minimal here.

Bioinformatics is thus, quite simply put again, technoscience born out of the convergence of the two main domains of application of cybernetics. It is *the* model of twenty-first-century technoscience: big yet distributed science, capital-intensive technology yet feasible in a garage, applied yet abstract knowledge at the level of its everyday practices. Out of the traditional disciplinary bounds, it is carried out at a fast pace in and out of academia, putting quite often the private sector and the university in competition for talents, patents, and finance. Both hyped discourse and down-to-earth straight capitalist business, it fuels our utopist dreams and dystopist nightmares alike.

Bioinformatics is the realm of technical experts of all kinds, starting with what technoscience had to contribute in terms of the least possible sociable "knowledge workers": entry-level obsessed computer programmers and lab-bench enslaved molecular biologists, sweating over their nth postdoc or residency, the new cyber-proletarian class. If they were paid by the hour, they would make less than your plumber. If they were financially rewarded for their diplomas, no company could afford to hire them (and a university even less so). They are the cannon fodder of capital's latest battles, the biotech battles. These are two kinds of employees who usually do not give a rat's ass about social *discourses*, unless they do not program or manipulate anymore—which usually comes only later in their careers, when they are half burned-out and ready for administrative duty. And why should they bother? Their domain of competence does not include such qualifications. Higher up in the food chain, or, better yet, in other *departments*, there are other highly qualified experts in these matters: lawyers, in-house bioethicists,[97] public-relations specialists, different kinds of doctors (and for the latter, *spin doctors*).

Bioinformaticians are very specialized workers laboring on high-demand, high-payoff, intensive schedules mostly composed of *routines*. If there is one *cliché* that goes against the grain today about these workers, we owe it to the desperate fiction of Michel Houellebecq: "research in molecular biology requires absolutely no creativity, no invention; it is in fact an activity almost completely routine-like, which requires only reasonable second-rate intellectual capacities"[98] (and programming might be worse, I might add). Bioinformaticians are technicians, that is, technical experts in the continuing age of the systems.[99]

This might still be a gross exaggeration by today's standards, but what about in ten or twenty years? For, just after his extraordinary claim, Houellebecq puts these words in the mouth of his fictional head of the biology department of the French Centre National de la Recherche Scientifique (CNRS): "To invent the genetic code . . . to discover the principle of the synthesis of the proteins, yes, one had to sweat over it a bit . . . But to decrypt DNA, pfff . . . one decrypts, and decrypts again. One does a molecule, then another. One inputs data in the computer, and it computes the subsequences. One sends a fax to Colorado: they decrypt gene B27, we take care of C33. It's like cooking. Once in a while there is an insignificant technical problem; usually it's enough to get you the Nobel Prize. It's tinkering; quite a joke."[100] Bioinformatics means the automation of lab work, the creation of enormous assembly lines devoted to number crunching expressed in sequenced and annotated base pairs (now counted in billions, Gbp). An intellectual activity with a taste for the *mystique* of the lonely discoverers, eccentric characters with macho bravado and bow ties, has first turned into a cottage industry and then into a global business run according to the good old principles of Taylorism. Delocalization has already started toward India . . .

Think about Britten, who told me that he had got into a new career by disgust for "big science," only to add in the same sentence, "well, the genome researchers have bigger lists of authors [on their published papers] now."[101]

Bioinformatics is, indeed, a capital-intensive activity. Billions of dollars of venture capital and public funding invested in the creation of thousands of companies, according to the seasoned logic of fixed capital: investments in infrastructures, machines, equipments, laboratories, top-notch facilities. The activity has moved out of arts and crafts in less than twenty years after getting out of its prehistory; nowadays, laboratories are regularly visited by the traveling salespersons of biotech companies pushing a new product, a new piece of equipment, a new piece of hardware or software. Any lab is run like an enterprise, whose main technicians, or even researchers, are often paid in "soft money." Performance rules; patent is king.

And what performances! In 2006, 434 genomes and 150 billions of base pairs were fully sequenced. From 1982 to the present, the number of bases in GeneBank has doubled every eighteen months, at approximately the rate anticipated by Moore's Law for computing, hence confirming the importance of computerization for the field of bioinformatics. And so many patents! In September 2005, 20 percent of the 23,688 human genes archived in the NCBI database were already tagged with at least one patent. Fourteen percent of the human genes so "patented" were held by private interests in the United States, with the company Incyte "owning" almost 10 percent of human genes.[102] "All your base are belong to us," might have been more prophetic than originally thought.[103]

It started slowly, however, at the rhythm of some key technical achievements of the late 1970s. Three main innovations are usually considered to be at the origins of the field: (1) recombinant DNA technique, (2) rapid DNA sequencing techniques, and (3) interactive computer programs allowing the handling of sequences produced by (1) and (2). All of these three main innovations were successfully developed during the 1970s, leading to complete automation of DNA sequencing in the 1980s. Paul Berg for (1), Frederick Sanger and Walter Gilbert for (2), earned the Nobel Prize in chemistry for these achievements in 1980. In his presentation speech for the three laureates at the 1980 Nobel reception, Bo G. Malmström called these achievements "methodological contributions," and added "the recombinant-DNA technique is . . . together with methods for the determination of nucleotide sequences, an extremely important tool to widen our understanding of the way in which the DNA molecule governs the chemical machinery of the cell."[104]

The first tool, the recombinant DNA technique, allows the production of an artificial DNA molecule by combining the DNA of two or more different organisms. Paul Berg created the first DNA recombinant molecule in 1972, using a restriction endonuclease[105] to isolate a gene from a human cancer-causing monkey virus and a ligase[106] enzyme to join the isolated gene with the DNA of the bacterial virus lambda. Stanley Norman Cohen and Herbert W. Boyer systematized the recombinant DNA technique in 1973, thus providing the key standard tool for genetic engineering.[107]

The second tool, DNA sequencing,[108] resulted from the work of the British biochemist Frederik Sanger (b. 1918) in the late 1960s. Sanger was already a Nobel laureate at that time, since he had earned the Nobel in chemistry in 1958 for his work in determining the complete amino-acid sequence of the bovine protein insulin in 1955. In the 1960s, Sanger turned his attention to RNA and, with his colleagues George G. Brownlee and Bart G. Barrell at Stanford University, developed a small-scale method for the fractionation of oligo-nucleotides that became the basis for most subsequent studies of RNA sequences.[109] This approach was slow and tedious, however, and the researchers realized soon that a new approach was needed. This led to the development of the "plus and minus technique,"[110] and later to the "chain termination technique"[111] (also known as the Sanger method) in the mid-1970s. At approximately the same time, Walter Gilbert (b. 1932) and his colleague Allan Maxam at Harvard University developed yet another sequencing technique based on chemical modification and subsequent cleavage of DNA at specific bases.[112] The Maxam-Gilbert method (or "chemical sequencing") became rapidly more popular than the "plus and minus" method, but was soon replaced in its turn by the more efficient chain-termination method. The Sanger method was still a highly manual process, however, with a cycle time of about one day and a throughput of only 100 base pairs (bp) per day. The development of this method led to the sequencing of the genome of the first viral organism in 1977, the Bacteriophage phi-X174.[113]

The first breakthroughs in sequencing techniques, and the resulting decryption of the entire genome of an organism, led in turn to the development of the third set of tools, the computerized tools that eventually gave the field its name at the end of the 1980s. The first programs were designed in the 1970s for naive users on interactive mini-computers and already provided eight main functionalities: "(1) storage and editing of a sequence, (2) producing copies of the sequence in various forms, e.g. in single or double stranded form, (3) translation into the amino acid sequence coded by the DNA sequence, (4) searching the sequence for any particular shorter sequences, e.g. restriction enzyme sites, (5) analysis of codon usage and base composition, (6) comparison of two sequences for homology, (7) locating regions of sequences which are complementary, and (8) translation of two sequences with the printout showing amino acid similarities."[114]

Following Sanger's decrypting breakthrough, many researchers attempted to develop full-automated sequencing. In the early 1980s, Leroy E. Hood (b. 1938), with his postdoctoral student Michael Hunkapiller and his colleague Susan Horvath, developed at the California Institute of Technology the first DNA synthesizer, a machine that strings together fragments of DNA to manufacture artificial DNA. Continuing their work at Caltech, Hood and Hunkapiller worked with Lloyd Smith, and later formed the company Applied Biosystems (ABI), which pioneered the automated DNA sequencer with its Prism370 machine in 1986. This machine was able to sequence DNA rapidly by labeling its four letters with laser-activated fluorescent dyes (rather than radioactive markers), and then automatically interprets the resulting DNA sequences. The Prism370 had a cycle time of about six hours and a daily throughput of 6,400bp, more than ten times the speed of the original Sanger method. Later improvements in capillary electrophoresis led to the first fully automated high throughput sequencer, the Prism3700 in 1997, with a throughput again ten times higher than the previous generation. In 2005, "the latest generation technology, the model 3730, with appropriate upstream automation, can spell out up to 2 million base pairs in the same time period, and at half the cost of the earlier model 3700."[115]

A resulting deluge of data suddenly fell over molecular biology during the 1990s. This wealth of data came from the inception of new techniques constitutive of the new field of bioinformatics. Between automation and libraries—of programming routines as well as sequence data—between new equipment and new products in kits, technology profoundly changed the field of molecular biology, to the point that some already saw there "an emerging new paradigm" for the field:

> Fifteen years ago, nobody could work out DNA sequences, today every molecular scientist does so and, five years from now, it will be purchased from an outside supplier . . . The new paradigm, now emerging, is that all the "genes" will be known (in the sense of being resident in databases available electronically), and that the starting point of a biological investigation will be theoretical . . . For 15 years, the DNA databases have grown by 60 percent a year, a factor of ten every five years . . . To use this flood of knowledge, which will pour across the computer

> networks of the world, biologists not only must become computer-literate, but also change their approach to the problem of understanding life.[116]

Gilbert, a pioneer in sequencing while it was still done by hand, had again anticipated another trend. At least he got it, albeit in both a more nuanced and authoritative fashion, a good seven years before Houellebecq did; and keep in mind too that he wrote this text two years before the actual birth of the World Wide Web! Today, what Gilbert feared or called for (it is not always clear through reading his text) has pretty much happened. For a part at least, *the biological understanding of life has become a software problem.* By this I do not mean that the understanding of life required the right software, but rather that life itself had become equated with a software problem. Once the program metaphor had gained a quasi-hegemonic status over the biological representations and that automation of data acquiring and annotating had reach its full potential, the field got caught in an almost too perfect convergence.

D **Hyperreal Junk**

> McLuhan's formula, *the medium is the message* . . . is the key formula of the era of simulation . . . This very formula must be envisaged at its limit, where, after all contents and messages have been volatilized in the medium, it is the medium itself which is volatilized as such . . . A single *model,* whose efficacy is *immediacy,* simultaneously generates the message, the medium, and the "real."[117]

During its bioinformatics revolution, biology has entered the new "real" world of simulation, or, to put it differently, biology has got to be less and less about wetware. Today, many contributions stemming from bioinformatics studies do not bother at all with experimental protocols dealing with living material. Models and graphs, equations and correlations, are slowly taking over bench work and observations, the days of the naturalists are long gone. This is precisely the first level of simulation. But at a deeper level, Jean Baudrillard had already recognized in 1981 the root of simulation in the very principle of biomolecular analysis:

> In the process of molecular control, which "goes" from the DNA nucleus to the "substance" that it "informs," there is no more traversing of an affect, of an energy, of a determination, of a message. "Order, signal, impulse, message": all these attempt to render the matter intelligible to us, but by analogy, retranscribing in terms of inscription, vector, decoding, a dimension of which we know nothing—it is no longer even a "dimension," or perhaps it is the fourth . . . In fact, this whole process only makes sense to us in the negative form. But nothing separates one pole from the other, the initial from the terminal: there is just

> a sort of contraction into each other, a fantastic telescoping, a collapsing of the two traditional poles into one another: an IMPLOSION—an absorption of the radiating model of causality, of the differential mode of determination, with its positive and negative electricity—an implosion of meaning. *That is where simulation begins.*[118]

This no more analogical thinking: DNA is the model medium of the new age of simulation, of capitalism of the fourth kind.[119] Baudrillard alludes here to the critique of teleonomy that we, and molecular biologists with us, have been wrestling with since the inception of a cybernetic molecular biology. Devoid of a semantic dimension, language here becomes inseparable from program, albeit as a self-referential grammar. Strange loops become the norm, Moebius rules (as in *The Matrix*). In such a frame, program, language, and medium collapse into one world, the world of simulation, the hyperreal. In this world, we are, and life is, expression of a program (and artificial life is a pleonasm). As in the TV ad, where two men shake hands and start emitting the mandatory polite phrases prescribed by the etiquette in the form of modem "handshaking" noises.

We are the only meaning remaining, and this does not amount to much. *We are worth our code, not worth a rush.*

In this model, DNA is medium and message, RAM and ROM, a hard disk and its contents, a symbolic text, whose words are combinations of four letters, and whose combinations of words produce the lines of code of a genetic program (a program without programmer). These words, these lines, are nothing but addresses in a symbolic register ruled by evolved conventions and equivalences. This language cannot be other than mathematical: the code of life is written in C++. So it is not merely when the appropriate technique became available that the Book of Life became a software problem; rather, it was a software problem from the get-go: no, dear Friedrich, there is only software, there is no more wetware. When the appropriate tools became available, the hyperreal world of simulation took over life itself. Not surprisingly, in the time it took to do so, computer interfaces and operative systems in turn evolved toward natural languages and graphic user interfaces. *It is once the world was enfolded in a global network of personal (albeit "pumped-up") computers, that the human being could be described as a genetic database.*

Present, past, and future have no meaning in a world ruled by numbers. Present is this address, like any other past or future address of a world *in real time:* just a question of probability. Space is the name of the illusion, which fathoms an address on the postal code of an infinite real estate. It is just a question of speed and acceleration, first- and second-order derivatives of space and time. And so much data, so many addresses, that our poor brain makes of us an obsolete computer. We are the ghosts in the Turing machine. Reason is but an archaic axiomatics, logic the name we give to our errors (transcription errors, that is). "The mutation will not be mental, but rather genetic."[120] Genetic, that is to say, computational.

Yes, but what about junk, then? Doesn't it contradict the whole notion of the genetic program? Is it not precisely this dark matter that refuses the measure of the coding performance?[121] Not at all, dear reader; it is the best remaining opportunity for data miners, the land of plenty for aspiring biostatisticians. Junk DNA is a dream come true for simulation-craving bioinformaticians: untied to the matter of life, these dirty proteins of wetware, it functions perfectly in the abstract space-time continuum where program, language, and medium collapse into one. It is the best candidate yet for the abstracted life that comes in numbers. Thank God almighty, there is plenty of it! It is the assurance of years of running time, of CPU activity: computers will be happy at last, there is life after the Human Genome Project.

A blessing in numbers, junk DNA is the *aufhebung* of code itself.

In the bioinformatics framework, it becomes the whole range of imaginary solutions to an absence of problem (in Baudrillardian terms, bioinformatics is thus an art form): junk as another code (a metacode, a supracode, a code of codes, a nested hierarchy of codes), junk as language, junk as medium, all in one. A blessing indeed . . .

The simulation of junk DNA might have officially begun in the "Research News" pages of *Science* in 1986[122] (but in fact it started long before that year). Six years after Orgel and Crick published their "Ultimate Parasite" paper in *Nature*, and five years after Baudrillard wrote *Simulacre et simulation*, alleged "hints of a language in junk DNA" were uncovered, again in the "Research News" pages of *Science*, under the heading "MATHEMATICAL BIOLOGY," in 1994.[123] That same year, in the same pages, "Mining Treasures from 'Junk DNA'" made the news.[124] A year later, *Nature*, this time, published a page titled "The Boom in Bioinformatics" in its "Employment Review—Genetics" section.[125] In 1997, in the same section, bioinformatics had already moved into a "post-genomics age."[126] Ten years later it is both a garage industry and a multibillion-dollar deal.

The first piece of news of interest here refers to a simulation carried out in 1985 at the University of California in San Diego and published in April 1986 in the *Proceedings of the National Academy of Sciences* (USA). Under the less sexy title "Multigene Families and Vestigial Sequences," William F. Loomis and Michael E. Gilpin reported on their simulative attempt to "address the question of why there is so much DNA in the world."[127] More precisely, they built their computer model in order to test the hypothesis of a random origin for the existence of "excess," that is, "selfish" (said the news), DNA. According to the news page, their conclusion was that it could indeed be the case: "Loomis and Gilpin indeed find that, given enough time, a simple gene will blossom into a genome that contains many genes, some of which are members of multigene families and all of which are embedded in a very large proportion of dispensable sequences."[128] Or, in the authors' own words: "we find that a genome will approach equilibrium with a fairly large amount of dispensable sequences and will carry a fluctuating number of copies of a given gene . . . Thus, the occurrence of multigene families as well as excess DNA is a consequence of random duplications and deletions.[129]

In other words yet, they simulated Susumu Ohno's hypothesis of "evolution by gene duplication" and its consequential existence of vast amount of "vestigial sequences" (i.e., "excess DNA," or "dispensable sequences" or "garbage DNA," or pseudogenes). Tomoko Ohta confirmed this interpretation in a paper published the following year in *Genetics,* where she compared her own model to Loomis and Gilpin's model as variations on the same kind of hypothesis.[130]

This connection is especially interesting because it relates the alleged "news" of the simulation of junk DNA to a much older trend congruent with Baudrillard's intuition. Tomoko Otha, like William Loomis, is actually a longtime contributor to the field who had proposed her hypothesis long before the emergence of bioinformatics.[131] Otha was a student of Motoo Kimura, whose work even predated Ohno's work: it is actually their work that provided Ohno with the key figure of his "So Much 'Junk' DNA in Our Genome" paper: "6% of our DNA base sequences is utilized as genes (Kimura and Ohta, 1971)."[132] Motoo Kimura (1924–94) might actually have produced the original hypothesis, central to his "neutral theory of molecular evolution,"[133] which predicts that at the molecular level, the large majority of genetic change is neutral with respect to natural selection—thus making genetic drift a primary factor in evolution.[134] As new experimental techniques and genetic knowledge became available, Kimura expanded the scope of the neutral theory and created mathematical methods for testing it against the available evidence. In 1973, his student Tomoko Ohta developed a more general version of the theory, the "nearly neutral theory," which accounts for high volumes of slightly deleterious mutations. Kimura and Ohta were thus pioneers in the field of computational biology, since, as James F. Crow, his longtime colleague, put it, "much of Kimura's early work turned out to be pre-adapted for use in the quantitative study of neutral evolution."[135]

So, in this piece of 1986 news that was no news, a simulation of junk DNA was reported, and, what a shock, they found that junk DNA could exist (hyperreally exist, that is). That's the hard way to put it. A more nuanced way runs as follows: "The availability of DNA sequence data after 1985 represents a significant turning point for the neutral theory and its tests. Using DNA sequence data, Martin Kreitman and others devised statistical tests that could statistically distinguish between neutrality and selection."[136]

So the piece of news was news after all: 1985 marks the moment when the availability of DNA sequences made it possible to test the predictive value of Kimura, Ohta—and thus Ohno's models. All the irony, and, I contend, the interest, of the piece of news in *Science* is that the work reported was actually a test performed not on these "actual" sequence data but on an artificial, computer-generated genome. In my perspective, the still ongoing debate about the relative success of these models is of little interest compared to the fact that junk DNA had then formally entered the hyperreal world of simulation. From then on, a thousand simulations would bloom, exactly as a thousand loops had bloomed, for simulation and feedback actually run on the same principle.

One of the instances of such a proliferation is very interesting for my purpose, because it introduced yet another metaphor in junk: "junk as language." The news report from the pages of *Science* in November 1994 actually anticipated the publication of a paper by a group of Boston researchers titled "Linguistic Features of Non-coding Sequences."[137] Rather than a simulation per se, this paper applied statistical analysis to thirty-seven sequences of eukaryotes, eukaryotic viruses, prokaryotes, and bacteriophages from GenBank Release 81.0 (February 15, 1994), plus two other bacteriophages and a 2.2 Mbp *C. elegans* sequence that just had been published at that time. The two statistical tests performed were Zipf analysis and Shannon redundancy analysis. They found that "non-coding sequences show two similar statistical properties to those of both and natural languages: (a) Zipf-like scaling behavior, and (b) a nonzero value of redundancy function $R(n)$." They thus concluded that "these results are consistent with the *possible* existence of one (or more than one) structured biological language(s) present in non-coding sequences."[138] There used to be natural and artificial languages, and suddenly a third category was born: "a biological language." And how could "a language" be "present in sequences"?

In contrast, the coding regions of the tested sequences failed both tests. "An expected result, Stanley says, because the 'language' of the genes lacks key features of ordinary languages. 'The coding part has no grammar—each triplet [of bases] corresponds to an amino acid [in a protein]. There's no higher structure to it.' In contrast, junk DNA's similarity to ordinary languages may imply that it carries different kinds of messages."[139] In the protocol of the study, a binary executable file of the Unix operating system and a random sequence of bits were used as control samples. The first, representing an "artificial language," passed the test, and the second failed. A Cambridge geneticist interviewed for the news page gave this laconic statement: "it sounds extraordinarily interesting . . . though it might be hard to tell what it means."[140]

Indeed. The publication attracted a lot of comments, many of which questioned the methodology or the analysis. In the issue dated March 11, 1996, of the *Physical Review Letters*, three comments were published along these lines. The first claimed that the Zipf analysis could not "distinguish language from power-law noise,"[141] the second "undercut speculative arguments based on Zipf's law and Shannon redundancy" by showing that the original paper's "observations may be simple consequences of unequal nucleotide frequencies,"[142] and the third simply stated that "this work ignores the long accepted view of the linguistic community that, even if correct, Zipf analysis provides no useful information about natural languages."[143] Other studies were published, criticizing or supporting the original publication, pretty much until now.[144] The final word on the question might have come earlier, though:

> The coding and non-coding regions of the DNA obey different statistical laws differently when analyzed in this linguistic fashion. That does not necessarily imply that junk DNA is a language and coding DNA is not. It does suggest that we might study DNA using the same statistical techniques that have been used to analyze language, and that if information is contained in the junk DNA, it is less likely to be in the form of a code and more likely in the form of a structured language.[145]

Whatever their disagreements might have been on the issue of some alleged "linguistic features" in junk DNA, almost all the contributors to this ongoing debate are actively engaged in statistical analysis, and have pondered about the statistical differences in coding and noncoding DNA. Pretty much everybody agrees that noncoding DNA exhibits long-range correlations and coding DNA does not. Actually, some even claim that this is no news at all, since this fact was "known at least since 1981," and that these statistical differences "are used even in routine methods for discriminating between them."[146]

So, in the end, all this might have been about a news item, which was not new, about some hints of language, which was not a language! The question that has been at stake during the whole debate (What do these statistical differences mean?) remains open, but in the process, a semantic and metaphoric field got enriched with some new items with a bright future in the hyperreal world of simulation. It is not by chance, then, that several fictions used the topic resources provided by this scheme. One imagines the Holy Qu'ran encoded in DNA,[147] another, which circulates endlessly on the Web,[148] the Holy Bible, of course, but my preference goes to a fantastic jewel of an ironic novelette written by communications professor and science-fiction writer Paul Levinson. Levinson imagines that junk DNA could encode

> a primordial copy protection scheme. A copy protection technique from Hell. DNA as ultimate shareware: *use* this little program to your heart's content, enjoy it, be fruitful and multiply with it, implement it—let it implement you—but don't copy the words without authorization. Not unlike many of our own computer programs—and books—really.[149]

So, of the six potential anagrams for our most intimate fiber, the one that alludes the best to its junkish nature might be NDA (for nondisclosure agreement, of course).

In between these science fictions and the fictions of science, some have followed the same hints that have inspired Eugene Stanley, and looked for other levels of code in junk DNA. This is the case of Jean-Claude Perez, a French bioinformatician and ex-IBM employee. Perez holds a "diplôme d'ingénieur du Conservatoire National des Arts et Métiers" and a PhD in mathematics and informatics. He started his research career in artificial intelligence and auto-organization while he was working for IBM during the 1980s. At the beginning of the 1990s, he left IBM to carry on his research in bioinformatics, leading to a (failing) startup called Genum at the beginning of this century. For close to the last fifteen years, Perez has practiced bioinformatics at the fringes of mainstream science, in a garage-like setup, pretty much a one-man operation. He too claims to have found proofs "a hidden language in DNA."

His "proofs" rely on some serious data mining, looking for "resonances" in DNA sequences. What kind of resonances is, at first, what put him into troubled waters. The first phase of his work, during the 1990s, focused on the Fibonacci[150] sequence of integers as an adequate descriptor of correlations in base frequencies in DNA sequences. According to him, "the thousands of letters TCAG of DNA 'tend' to harmoniously auto-organize themselves by optimizing their relative proportions according to the Golden Ratio and the integers of the Fibonacci sequence."[151]

Fibonacci numbers have numerous applications in mathematics, and are known to represent adequately various natural structures such as shellfish patterns or flowering spirals. Luc Montagnier, the codiscoverer of HIV, even wrote: "Jean-Claude Perez's discovery interests me. It is very possible, indeed, that beyond the sequences, a kind of DNA 'supracode' exists. It could eventually explain these non-coding DNA sequences, which do not code for the synthesis of proteins, and which are considered useless. The idea of a law based on a mathematical sequence present within the heart of DNA has nothing surprising in itself. I find it quite natural, in fact."[152] Perez's book actually got two titles: "DNA Decrypted" on the cover and "The Supra-Code of DNA" on the first page. This "supracode," however, appears in the book as elusive as the key notion Perez uses to describe it: the golden ratio (φ). Two quantities (a, b; a>b) are said to be in the golden ratio if the ratio between the sum of those quantities (a+b) and the larger one (a) is the same as the ratio between the larger one (a) and the smaller (b), which is represented by the following equation: $\frac{a+b}{a} = \frac{a}{b}$, whose only solution is the irrational number $\varphi = \frac{1+\sqrt{5}}{2}$.

The golden ratio is approximately 1.6180339887. The relationship between the Fibonacci sequence and the golden ratio is as follows: the ratio of a number in the Fibonacci series over the previous becomes increasingly closer to the golden ratio as the Fibonacci numbers increase (the ratio of two successive Fibonacci numbers approaches phi at the limit). The golden ratio has attracted artists and mathematicians alike since there have been arts and mathematics,[153] but some consider that its presence in discourse is still the sure mark of some deranged mind.[154] This second perception was not lost on Jean-Claude Perez, who told me that "there was another aspect that explains why I changed the focus of my research: the golden ratio is a dead end; nobody would ever accept a discovery featuring the golden ratio, even if it were true."[155] In a second phase of his research, Perez thus moved to different forms of data mining in DNA sequences, and found many other levels of codes.

Today, he claims that he has found at least eight levels of codes, and makes no more mention of the Fibonacci sequence or the golden ratio. He first did some number crunching not on the sequences themselves, but on the sequences of the bases characterized by their atomic masses. He found that there always exists a "perfect balance" in the atomic masses of each of two complementary DNA strands. This seemed to me to be a trivial result following Chargaff's parity rules.[156] When I asked him, he answered something quite incomprehensible, only to point me later to an unpublished paper with the following answer: "In fact, the laws of Chargaff about the numbers of TCAG bases are a consequence of this balance of the masses . . . It thus results quasi a balance between related masses of both strands. But this balance then will be adjusted and amplified by the exact relations between the atomic masses of bases TCAG."

This still seems incomprehensible to me, and, moreover, contradictory; it claims at the same time that the discovery does not revisit Chargaff's rule and that Chargaff's rule is a consequence of the discovery. But since the study relies on computation alone, it cannot pretend to explain the relationship between a statistical observation and a chemical observation. In other words, all seem to hold in the impossible demonstration of a pairing of bases *in order to* "compensate" masses for complementary DNA strands "equilibrium" (the return of the teleological *specter*). The statistical observation might be accurate, but the logic seems flawed. Anyway, the literature is already quite clear about this. Contrary to his first parity rule, Chargaff's second rule was discovered late,[157] and although "scientists have gathered very strong evidence for its general validity," it still has no "generally accepted explanation" and "continues to stimulate the search for their unknown underlying mechanism."[158] A duo of Brazilian researchers have actually used the second parity rule as a postulate to build a mathematical model that "established an interesting connection between nucleotide frequencies in human single-stranded DNA and the famous Fibonacci numbers"![159] Proof is that Perez could have intuited and confirmed alone the results of data-mining and statistical-analysis studies of the genome. This does not mean necessarily, however, that his understanding of the biological aspect of the question was flawless. Actually, the rest of the story seems to confirm the opposite alternative.

Perez then attempted to further decode DNA on the basis of other statistical analyses on sequences and atomic mass data. He claims that this work led to the discovery of a set of "universal numeric codes" unifying all genetic information, from the atomic level to the level of whole genomes. These codes are interdependent and embedded, and testify to some kind of self-organization, since Perez's analyses demonstrated their fractal nature (self-similarity and scale-independence). According to him, the strangest phenomenon he discovered relates to "the obviousness of 'discrete' undulatory waves," on the one hand, and the emergence of a universal, yet self-emerging, "binary code," on the other hand. Details of these eight codes and their biological significance are not clear to me to this day. But whatever their meaning might be, I think that Perez gave me a clue to his framework when he told me, "One can look for any structure in DNA and find it."[160] Retrospectively explaining his "supracode," he concluded:

> The supracode . . . is a real law, but it is not a code in itself, in the sense code is usually understood. It is a bit like if in a language the message was at the same time semantics, rhetorics, even tone and silence, code and all. This order thus seems to be the consequence of other, deeper codes.

No wonder I found there a perfect illustration of hyperreal junk. But then again, maybe junk DNA, like whole DNA in fact, was hyperreal from the start, when the analogy between linguistics and genetics was first proposed.

Chapter 3, Head Again

Multimedium, or Life as an Interface Problem

> The electronic universe, with its model worlds and computer simulations, with its interfaces and virtual realities, presents strong evidence to support the belief that comprehension of the world really is an interface problem.
>
> —PETER WEIBEL, "The World as Interface"

Once medium, language, and message had collapsed into simulation, the performativity of the founding trope of the genetic program was bound to proliferate in many uncontrolled ways. Once life had become a software problem, and the program an auto-organized production, the linearity of the archetype gone in myriads of networked effects, room was left for the ekstasis of communication. Passed this line, one encounters speculations, ramified, paradoxical, contradictory speculations. The time has come for theorizing without a "solid factual base," the time of soft theories, versus solid, material, empirical theories; evidences are now declared outlawed. A metaphor has run amok, in an endless process mirroring the power of the wandering mind alone.

Junk DNA redoubled in junk theories, red-herring analogies, untestable hypotheses, overstretched notions, and unfathomable processes: there is no more wetware, my dear Mobius. How is life above and beyond the fourth dimension? What do you mean by life as the emergent behavior of a quantum computer, or, better yet, of a fuzzy network of quantum computers? What do you mean, my soul? We are the medium: media "r" us. Welcome to the next dimension.

Medium: something in the middle: a means of effecting or conveying something: a surrounding or enveloping substance: the tenuous material that exists outside of large agglomerations of matter: a channel or system of communication: a mode of expression: something on or in which information may be stored: something held to be a channel of communication between the earthly world and the world of spirits: a go-between: a condition or environment in which something may function or flourish: a nutrient system for the artificial cultivation of cells or organisms: a fluid or solid in which organic structures are placed (as for preservation or mounting): intermediate in quantity, quality, position, size, or degree: something commonly accepted in exchange for goods and services as representing a standard of value: us, in all of our senses (all according to Merriam-Webster).

Our DNA, in these times of genohype, is this fetish, this surface of projection for all of these meanings. Remember, if you still can, there is no more meaning, albeit we are the only meaning left. We are left to be these subjects who might affect and be affected (one can still dream). We are but stations on the fluxes of circulating affect. *Circulation:* orderly movement through a circuit: flow: passage or transmission from person to person or place to place: the interchange of currency: the extent of dissemination: the average number of copies of a publication sold over a given period: the total numbers of items borrowed from a library: us, circulation: a few billion, and counting.

Our DNA, in these times of genohype, is this medium of an uncertain circulation. What flows, what is exchanged, what disseminates in there? What forms does this circulation take? Who could, at the time of the Web 2.0, be satisfied with a single modality, a simplex medium? Comes the time of the next great *déphasage:* from solid-liquid-gaseous to matter-energy-information.

Why should this machine be confined to be a chemical machine, and very inert at that? How about electricity, light, quantum states, streams of consciousness? Yes, how about that? How about action at a distance, action in time, delayed or anticipated, how about action in parallel universes? Yes, how about that? Mind over matter, sir Machine, comes the time of the multimedium!

No place for systematic thinking here: let me abandon the disguise of petty historical narrative, dates and places, firsts and lasts: chaos rules, neither God nor Genes. At best, there might be something like a fuzzy configuration emerging among the marginal thinkers, the kooks and cranks of revolutionary science—in my delirium about this delirious world, I once called it *antescience,* in between fictions of science and science fictions. So, be warned, dear readers: this prehistory achieved, one would be in his right mind to believe that we now live in the miracle of the antescience,[1] whose basic axioms may be:

- All you think you know about DNA is junk: the standard model was very useful for a while, but it now risks obstructing the truth, if it still claims that's all there is to it (proteins, matter, mechanisms).
- There is no such thing as coding DNA, if code relates only to protein synthesis (there is no gene).
- Conversely, there is no such thing as noncoding DNA; there is only DNA, in various forms during the life cycle of the cell.

- The fact that one or various codes, one or various languages, one or various messages implicate DNA or parts of DNA is open to speculation.
- There is more to reality, life, the world, than you can directly experience.

E **The Field**

Here I start by returning on Jeremy Narby's steps, and before him, those of his teachers, the McKenna brothers. They were the central inspiration behind his intuition of a direct communication between the entranced shaman and the DNA of the world. They first intuited a molecular mechanism (intercalation) linking a hallucinogenic molecule (DMT) and the *genic* molecule (DNA).[2] Moreover, they gave precise indications of a broader speculative model that links nucleic acids and consciousness via some quantum processes involving electromagnetic waveforms emission and reception in synapses.[3]

For Narby, these experiences are akin to audiovisual processes occurring during the shamanic trance: he calls them, following the shaman who had initiated him, "the television of the forest."[4] It literally means that, during the trance, the shaman sees and hears an otherwise mute and invisible reality: the use of the hallucinogenic substance, as well as some other "technics of ecstasy" (Mircea Eliade), allows him to access this reality. For Narby, the crucial mediation is that of DNA: "in their visions, shamans take their consciousness down to the molecular level and gain access to information-related DNA, which they call 'animate essences' or 'spirits.'"[5] The tobacco and other hallucinogens used by the shaman attract these craving spirits, called *maninkari* (literally, those who are hidden). Here is Carlos Perez Shuma's (Narby's informer) account of the process:

JEREMY: You told me that ayahuasca and tobacco both contain God.
CARLOS: That's it.
JEREMY: And you said that souls like tobacco, why?
CARLOS: Tobacco has its method, its strength. It attracts the maninkari. It is the best contact for the life of a human being.
JEREMY: And these souls, what are they like?
CARLOS: I know that any living soul, or any dead one, is like those radio waves flying around in the air . . . That means that you do not see them, but they are here, like radio waves. Once you turn on the radio, you can pick them up. It's like that with souls; with ayahuasca and tobacco, you can see them and hear them.[6]

Here is, I guess, one of the possible formulations of the "revolution" I am talking about: DNA, in this narrative, finds its renewed analogy out of the culture of print that gave us the text-program metaphor, into the electromagnetic wave broadcast culture. Hence, DNA becomes a modern medium. Since the coding part of DNA had been successfully proved to be the medium of the physical mechanisms of life and heredity, its noncoding part would become, in this new framework, the medium of its immaterial part. DNA thus becomes a multimedium.

Narby narrowed his investigation to the literal meaning of his informer's discourse, and focused on the audiovisual aspects of the manifestations. Thanks to the work of Fritz-Albert Popp, a German biophysicist, he was able to link them to DNA photonic emissions (biophotons): "during my readings, I learned with astonishment that the wavelength at which DNA emits these biophotons corresponds exactly to the narrow band of *visible light*... It was almost too good to be true. DNA's highly coherent photon emission accounted for the luminescence of hallucinatory images, as well as their three-dimensional, or holographic, aspect."[7] He went further, and assigned a "possible new function for junk DNA" in this process: "I suggest the hypothesis that DNA's 'non-coding' repeat sequences serve, among other things, to pick up photons at different frequencies."[8]

I decided to go to the International Institute of Biophysics at Neuss, near Düsseldorf, and interview Fritz-Albert Popp. While preparing for my interview, I found his work amply quoted and discussed in a book that attracted my attention for the first time on the "revolutionary synthesis" that I am attempting to discuss here: *The Field: The Quest for the Secret Force of the Universe,* by an investigative journalist named Lynne McTaggart.[9] In the prologue to her book, she describes a loosely connected network of "dissatisfied" scientists who, she claims, rediscovered the repressed equations of quantum physics about the "Zero Point Field" (or simply "The Field" hereafter), "an ocean of microscopic vibrations in the space between things... one vast quantum field... connect[ing] everything... like some invisible web":

> On our most fundamental level, living beings, including human beings, were packets of quantum energy constantly exchanging information with this inexhaustible energy sea. Living things emitted a weak radiation, and this was the most crucial aspect of biological processes. Information about all aspects of life, from cellular communication to the vast array of controls of DNA, was relayed through an information exchange on the quantum level . . . We literally resonated with our world. Their discoveries were extraordinary and heretical.[10]

She portrayed Popp as one of these dissatisfied scientists in her third chapter, titled "Beings of Light," and credited him with the discovery of biophotons in an experiment with ethidium bromide, a chemical that unwinds the double spiral of DNA. With an analogy, she explained that, according to Popp, "one of the most essential stores of light and sources of biophoton emissions was DNA. DNA must be like the master tuning fork of the body."[11]

I asked Popp if he agreed with this narrative, and he answered that he did not completely agree with it, that there was "a little science fiction in it," since this work was still speculative at the time, and did not constitute a proof.[12] I had also read on the Web that Popp's work permitted one to describe "a virtual electromagnetic field that pervades the entire organism with a virtual photonic flux. In this field, virtual photons are stored. The field continually receives inputs from the environment and is continually outputting biophotons, particularly in the near ultraviolet."[13] So I also asked him about this, and again he told me that he agreed with the idea of the existence of "some kind of electromagnetic potential field."

I questioned him then about the part that DNA, and especially its noncoding part, was playing in this phenomenon. He first answered that he thought that photonic storage and emission involved DNA as a whole, but he soon nuanced this answer in taking into account the differences of activity and structure between the coding and the noncoding part of DNA. According to him, the presence of repetitive sequences in noncoding DNA implied a higher regularity and therefore a higher capacity for virtual photons storage for this part. He also insisted on the differential speeds involved in the cellular life: when the coding part of DNA is principally involved in the chemical synthesis of proteins and operates at relatively low speed but in huge numbers (in the order of hundreds of thousands reactions per second), whole DNA, including junk, regulates all this activity thanks to a communication network working at the speed of light. He insisted again that this was still speculative thinking, even if he found it "necessary that it works this way." Overall, he thus confirmed Narby's intuitions and generally agreed with a global picture of the whole process as depicted in McTaggart's synthesis. The following excerpt from an article published in the journal of the Institute of Noetic Sciences summarizes this "big picture":

> Suppose that DNA is using electromagnetic radiation, coherent light, to communicate inside the cell . . . Could it be that the cell uses a two-level communication system—slower-speed communication via conventional molecular processes that take place around the DNA molecule and a much-higher-speed communication within the whole cell using coherent light? . . . The entire organism may be a complex flow of information in which each cell and organ is responding to a constant flux of electromagnetic messages.[14]

This coincidence is not hazardous, since the Institute of Noetic Sciences (IONS) is one key institution in the international network devoted to the advancement of the ideas of the new synthesis described here. Former Apollo 14 astronaut Edgar Mitchell founded IONS in 1973 to explore "the new frontiers of consciousness" (*nous* or *noos* means consciousness, mind, intellect, or intuitive knowing in Greek). It is a nonprofit voluntary organization with approximately thirty-five thousand members worldwide, headquartered in Petaluma, California. A former director of the institute defined the "noetic sciences" as "the esoteric core of all the world's religions, East and West, ancient and modern, becoming exoteric, 'going public,'" and quoted Aldous Huxley's description of this "perennial wisdom": "it recognizes a divine Reality substantial to the world of things and lives and minds . . . finds in the soul something similar to, or even identical with, divine Reality . . . places man's final end in the knowledge of the immanent and transcendent God of all Being."[15]

According to its vision and mission statement, IONS "explores the frontiers of consciousness, builds bridges between science and spirit, researches subtle energies and the powers of healing, inquires into the science of love, forgiveness, and gratitude, studies the effects of conscious and compassionate intention, seeks to understand the basis of prevailing worldviews, practices freedom of thought and freedom of spirit."[16] No wonder, then, that it is often described as a typical "new-age" institution by its detractors, such as Quackwatch (which put it on its list of "questionable organizations") or the Committee for Skeptical Inquiry.

Another "dissatisfied researcher," who benefited from research grants from IONS, was also on my list of interviewees: Rupert Sheldrake. Fritz-Albert Popp told me that he thought that Sheldrake was "essentially right" even if he had been "very skeptical" about his ideas at first. Many still are, to this day. Negative reviews from "serious scientists" have accompanied Sheldrake's publications since his first book came out in 1981.[17] *Nature* opened fire with an editorial rhetorically titled "A Book for Burning?" in which Sir John Maddox called the book "an exercise in pseudo-science" that "should not be burned . . . but, rather, put firmly in its place among the literature of intellectual aberrations."[18] *Nature,* however, published many responses from readers in its subsequent issues, all alarmed about the "harsh words" of the editor, his way "to obstruct this avenue of progress," etc.[19] His friend and coauthor Terrence McKenna went one step further when he commented on Sheldrake's work: "the overturning of a scientific paradigm is a political act, and it has to do with reputations, and tenure, and publication, and people who have built their lives defending something that they now see under severe attack."[20]

My guess is that Sheldrake (like Popp, in a lesser way)[21] especially infuriated the "serious scientists" like Sir John Maddox because he was, at first, one of them. He studied biochemistry at the prestigious University of Cambridge Clare College, graduating with a double First Class honors degree, and taught there for a while. He was a Frank Knox fellow at Harvard University and a research fellow of the British Royal Society. But everything took a different turn in his life and career in 1974 when he went to work in crop physiology at the International Crops Research Institute for the Semi-Arid Tropics (ICRISAT) in Hyderabad, India. He lived there until 1985. During that time, he met the Indian philosopher Krishnamurti and spent a year and a half in Bede Griffiths's ashram, where he wrote *A New Science of Life.*

Sheldrake, indeed, became dissatisfied with the mechanistic worldview of molecular biology, and even more so with its metaphor of the "genetic program." In 1981, he criticized this notion for its tautological nature, a critique that we have already encountered (in chapter 1): "in so far as mechanistic explanations depend on teleological concepts such as genetic programs or genetic instructions, goal-directedness can be explained only because it has already been smuggled in."[22] In his next book, seven years later, the critique had grown even tougher: "the central paradigm of modern biology has in effect become a kind of genetic vitalism," he wrote, adding, "although nominally mechanistic, [it] has in effect become remarkably similar to vitalism, with 'genetic programs' or 'information' or 'instructions' or 'messages' playing the roles formerly attributed to vital factors."[23] In our interview, twenty-two years after his inaugural publication, he still maintained the same argument, equating the use of the genetic program metaphor to the "atomistic reductionist approach dominant in biology," leading to an improper focus on DNA:

> It's like saying I want to understand a building by understanding the ink on the architect's plan or the bricks in the building. It missed the point that DNA is always arranged in chromosomes and these integral structures of which every bit of DNA is a part. In a way molecular biology displays and extremely narrow reductionist focus and ignores the chromosomes; these structures that you actually see under microscopes and within which DNA is enclosed, included.[24]

"Reductionist," "mechanistic," and "atomistic" are three epithets that have been quite often used by the detractors—and especially the "new-age" detractors—of "mainstream science." On the opposite side, mainstream scientists often argue that most biological pseudoscience derives philosophically from vitalism,[25] "the doctrine that the functions of a living organism are due to a vital principle distinct from physicochemical forces" or, alternatively, "the doctrine that the processes of life are not explicable by the laws of physics and chemistry alone and that life is in some parts self-determining" (Merriam-Webster). "New age," say they, is a resurgence of vitalism, this long-abandoned doctrine.

Now, and this is harsh, molecular biology is so much vitalist, answers Sheldrake: the genetic program is its "vital principle." It just happens to be a *material* vital principle. I hope that you get the irony, and thus, the rage it must have created. It is, after all, returning molecular biology to the dead end of its "flagrant *epistemological contradiction,*" which Jacques Monod and the likes were desperately trying to resolve with their rhetorical reinvention of teleology under teleonomy's guise.[26] It sounded like, in fact, sheer mockery in the face of molecular biology to tell her that her precious "program" was no less "a vital principle." It seemed to forget that the "vital factors" had to be distinct from physicochemical forces; or did it? Did it mean, in fact, that genes, and most crucially their expression and regulation, could not be completely determined by physical or chemical factors alone? This is where the question of the epigenetic processes once again took center stage.

Conrad Hal Waddington (1905–75) coined the term "epigenetic landscape" in 1942, as a metaphor for biological development.[27] This neologism was based on the use of the Greek prefix *epi-* meaning "on top of" or "in addition to." Epigenetics thus originally referred to the processes of biological development by which genotype gives rise to phenotype, with no connotation of the nature of the said processes.[28] Today, it usually refers to the processes of heredity that do not involve changes to the underlying DNA sequences. Waddington later refined his concept with yet another Greek neologism, the notion of "chreode" or "development pathway" ("chreode" is derived from the Greek words *chre,* necessary, and *hodos,* way or path).[29] Pursuing his original spatial metaphor, Waddington envisioned these epigenetic pathways as the valleys and ridges of a mountainous landscape canalizing the movement of a ball (see Figure 7).

The central notion here is the necessity of the pathways as predetermined avenues for development. This determinism, however, differs from genetic determinism *stricto sensu.* In fact, it attempts to explain how the same genes could be expressed differently in development, leading to differentiated cells, tissues, and organs. We have seen previously that the standard explanation today considers this as a problem of regulation of the genetic expression and includes such processes as histone modification, methylation, gene silencing, and so on.

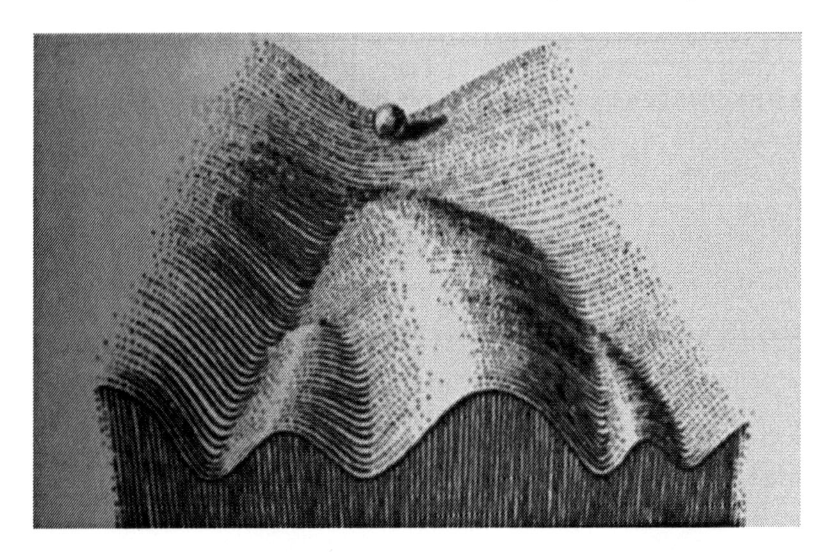

Figure 7. Waddington's original representation of the chreode. From *The Strategy of Genes* (London: George Allen and Unwin, 1957).

Waddington's original idea of epigenetic processes depended on the notion of the "individuation field." The spatial metaphors of the epigenetic landscape and the chreode were thus "descriptive conveniences," the means to represent some immaterial or nonmaterial influences on development. This is exactly where Sheldrake followed his lead. He made the hypothesis that Waddington's individuation fields could be considered "morphogenetic fields," real as any known physical field, but determined only by the past shapes and forms taken by former similar individuals. He called this hypothesis alternatively "formative causation" or "morphic resonance":

> These fields order the systems with which they are associated by affecting events which, from an energetic point of view, appear to be indeterminate or probabilistic; they impose patterned restrictions on the energetically possible outcomes of physical processes . . . they are derived from the morphogenetic fields associated with previous similar systems: the morphogenetic fields of all past systems become *present* to any subsequent similar system; the structures of past systems affect subsequent similar systems by a cumulative influence which acts across both space *and* time. The hypothesis is concerned with the *repetition* of forms and patterns of organization; the question of the *origin* of these forms and pattern lies outside its scope.[30]

The notion of "resonance" serves as a catachresis here; "energetic resonance" appeared to Sheldrake as "the most appropriate physical analogy" to describe the modus operandi of morphogenetic fields, these unknown entities. Resonance is the event that occurs "when a system is acted upon by an alternating form which coincides with its natural frequency of vibration."[31] Examples of such phenomena are the vibration of a stretched string of a musical instrument, the tuning of a radio set, the absorption spectra of atoms, or Electron Spin Resonance. Unlike energetic resonance, however, morphic resonance does not involve transfers of energy and depends on four-dimensional patterns of vibration (in space-time). The introduction of notions of vibration, transmission, and reception thus placed Sheldrake's hypothesis clearly into the mediatic domain of reference, a point that did not escape his attention:

> This raises the problem of the medium of transmission: how does morphic resonance take place through or across time and space? In answer to this question, we might imagine a "morphogenetic aether," or another "dimension," or influences passing "beyond" space-time and then re-entering. But a more satisfactory approach might be to think of the past as pressed up, as it were, against the present, and as potentially present everywhere. The morphic influences of past organisms may simply be present to subsequent similar organisms.[32]

One might be tempted here to translate Sheldrake's hypothesis in the Aristotelian vocabulary: the past forms exert a virtual influence on the present form, which actualizes them through differences and repetitions. This Aristotelian resonance was very clear to Sheldrake, who discussed at length the differences and similarities between his conception of the morphogenetic field and embryologist Hans Driesch's recycling of the Aristotelian concept of *entelechy*. Stemming from the Greek words *enteles* (complete), *telos* (end, purpose, completion), and *echein* (to have), Aristotle's *entelecheia* refers to a force, principle, or agency that exists formally within something and guides the actualization of its potential toward full being. Entelechy is thus the formal or formative cause that actualizes the potential of being, and as such it names the force or principle that leads to self-fulfillment. In the vitalist theory of Hans Driesch (1867–1941), the entelechy was thus the new name of this non-physicochemical vital principle, this inherent agency that regulates and directs the development and functioning of an organism.

Like genetic programs and entelechies, Sheldrake's morphogenetic fields and morphic resonance are a theoretical solution to the problem of teleology in heredity and development, and allude to some kind of self-organization. Like the genetic program, the morphogenetic fields can be seen as information structures, fields of information.[33] Unlike the genetic program, they propose a nonmechanistic picture of these phenomena, and refer to processes external to the genetic material. Although they share this characteristic with Dreisch's entelechy, they differ from it in their analogy with the physical fields and thus avoid its vitalism.

Sheldrake noted, however, that "if morphogenetic fields are considered to be fully explicable in terms of known physical principles, they represent nothing but an ambiguous terminology superimposed upon a sophisticated version of the mechanistic theory. Only if they are assumed to play a causal role, at present unrecognized by physics, can a testable theory be developed."[34] When I asked Sheldrake what he thought of this particular point more than twenty years after his initial formulation of the morphic resonance hypothesis, here was his answer:

RS: In my idea about morphic resonance I think that there are fields organizing systems at all levels of complexity: molecules, organelles, cells, tissues, organs, so I think that morphic resonance and morphic fields would apply to DNA molecules and to entire chromosomes.

TB: Twice, when you talked about those levels, you started at the molecular level, is there something under it?

RS: Well, I think that atoms have fields; those are described by quantum field theory. I think that quantum fields and what I call morphic fields are rather similar. I wouldn't like to say that morphic fields are quantum fields because they are dealing with systems at higher levels of complexity than quantum theory has described. But I think they're more similar to quantum fields than they are to electromagnetic fields or other fields.[35]

At no point during our interview, however, did Sheldrake ever mention the Zero Point Field. The closest reference I found in the literature came from his discussion with physicist David Bohm about "morphogenetic fields and the implicate order."[36] First, Bohm was prompt to remind Sheldrake that quantum theory does not have any concept of continuity in time, or process, or even actuality. He then proposed his own reformulations of the theory, which, according to the editor of the dialogue, "are rather similar to Sheldrake's theories."[37] These reformulations allude to Bohm's two main contributions, from the quantum potential to the implicate order:

> The set of particles, the whole structure of all the particles forming a system, is the actuality of that formative field. But that model by itself still ignores time, so the next step is to bring in time, to say that there's a succession of moments

of time in which there is a *recurrent* actuality. And we would say that what recurs is affected by the formative field. But then that formative field is affected by what has previously happened, actually. *Now that would help to remove most of the problems in physics, if we can manage it.* And it would tie up closely with the sort of things that you're talking about.[38]

I do not have the time, the energy, and, quite frankly, the skills to get deeper in this matter here.[39] So I will content myself with the following proposition: Lynne McTaggart's thrice-repeated allusion to Bohm's implicate order[40] and the coherence of his conception with this new-age synthesis allow us to consider that morphogenetic fields, formative causation, and morphic resonance depend indeed on some not yet recognized or understood aspects of quantum fields, possibly anticipated in Bohm's interpretation of quantum mechanics. McTaggart would seem to agree with this proposition since she wrote: "Sheldrake's theory is beautifully and simply worked out. Nevertheless, by his own admission, it doesn't explain the physics of how this might be possible, or how all these fields might store this information. In biophoton emissions, Popp believed that he had an answer to the question of morphogenesis as well as '*gestaltbildung*'—cell coordination and communication—which could only occur in a holistic system, with one central orchestrator."[41] When I asked Sheldrake about this convergence, however, he was quite dismissive and answered: "Biophotons are a reflection of the coherent organization of living organisms . . . they are an effect of that organization, rather than the cause."[42]

Whatever the case, for my purpose, Sheldrake, like Popp and Narby, provides an alternative metaphoric realm to that of the "genetic program." It is the realm of the multimedium, with its broadcasting references. Sheldrake, indeed, recurs frequently to radio and TV metaphors to explain his notion of morphic resonance. Here is one instance of such metaphors in his work:

> The music which comes out of the loudspeaker of a radio set depends *both* on the material structures of the set and the energy that powers it *and* on the transmission to which the set is tuned . . . In terms of the hypothesis of formative causation, the "transmission" would come from previous similar systems, and its "reception" would depend on the detailed structure and organization of the receiving system.[43]

In our interview he confirmed these metaphoric uses, but insisted on their limits. Here are his answers when I asked him whether "antenna" would be an appropriate metaphor to describe the function of certain cellular structures (including DNA):

> I don't usually use the word *antenna*, I use the word *resonance*. Antenna implies a particular metaphor, namely, radio, or television, whereas resonance is a more general metaphor, which applies to radio and television but also applies to acoustical resonance like that of a piano string. So, I use the metaphor resonance, which is much more general, and occasionally when people are trying to understand or force me to give a metaphor, then I do use the radio metaphor for several reasons. First, I use it to give the idea of a resonant tuning. That's the antenna aspect. Second, the fact that you can change some components of the radio or TV sets and distort the reception, but in neither case does it prove that everything [the signal] is stored inside it.[44]

Although Sheldrake refuses in his holistic view to ascribe a given metaphor to junk DNA in particular, but rather to whole chromosomes, and only as one level among many other nested levels of morphic resonance, we can see here an instance of a fundamental metaphor: that of a transmitter-receiver set, of a possible quantum nature. In the age of Web 2.0, the metaphors of the radio or the television might appear quite passé. Moreover, they are inappropriate for my purpose here, because they rely on a broadcast system, a form of one-to-many communication system. In Sheldrake, Narby, or Popp theories, communication indeed appears to be a many-to-many system participating in a one-to-many system. This is one more reason to consider the multimedium model as the most efficient for descriptive purposes. Not only could DNA be described as a multimedium if the chemical communication of protein synthesis was complemented by other communicative modalities such as biophotons emission, but the nature of the considered interactions between many distributed agencies would make this analogy even better.

This is why the networked computers of today's hypermediatic systems will eventually appear as the best metaphor to renew the tired cliché of the genetic program. If the computer becomes a quantum computer, and the communication wireless, the analogy gets even better: in short, *a bio-quantum Web.* This is Gaia and its noosphere all in one; talk about an era of convergence! The World Wide Web, of course, already displays this reflexive analogy:

> Our DNA is able to create magnetized wormholes through these they can communicate with higher dimensions. This communication is non-linear (meaning it does not follow our logical understanding and reasoning) and is immediate, thus space-time free. This process is also called hypercommunication and it is through a network of different levels of consciousness. This communication does not serve a special, limited purpose but just like the Internet the DNA put their own data into this higher dimensional network, can call upon data from this network and also establish direct contact to another member of this network.

> The DNA can therefore have its own homepage, can surf in the higher dimensional network and can also chat with other members.[45]

This "hypercommunication" actually refers to Popp's and Sheldrake's theories, but also to those of Matti Pitkänen, a theoretical physicist from Finland, and those of Pjotr Garjajev, a Russian biophysicist, pioneer of the "biological Internet." In spite of some desperate attempts, I must confess that I do not feel that I can adequately describe their researches.[46]

I went to Hanko, Finland, to interview Matti Pitkänen during the summer of 2003. I spent four days with him, enough to get to know him a bit. I came back with the definite feeling that I had lived through some sort of awakened dream. Some words still flashed in my mind . . . p-adic numbers . . . many-sheeted DNA . . . mirror neurons . . . but no clear picture of how I could ever make sense of this. It might have been the most difficult interview I have ever done. Its transcript is full of holes; Matti was barely audible most of the time. I met some of his friends, strolled in the streets of his city by the sea, and watched the trains come and go. At night in the little room near his own, I might have dreamt I was there.

There, I felt things I could not express. Everything measured up to a unique criterion of coherence: internal coherence, as in the inversed mirror of the theory, but coherence with the world, too. Unfathomable numbers, discrete events in the fabric of a disjointed time . . . the quality of this negative relation that nevertheless touches upon the essential nature of the world: that which cannot be divided. Integers, primes, powers of primes, all resonating to some mystical echoes. I, too, am what cannot be divided. I am that quantum consciousness. Am I? And he, in his quasi-autistic fortress, still remembering his "episodes," social uneasiness, hypersensitivity, even possession . . . For him, we are this constant interfacing between fluxes of energy, negative energy in tension toward the past, positive energy reflected back from the future. We, the world, as in Narby's speculations: all is minded. All is connected. No wonder I had such a hard time, once back home, to navigate in the hundreds of Web pages devoted to some kind of new-age healings, doctrines, revelations. I gave up. I yielded to the powers of Maya. OK, (I am this bio-quantum) computer. I compute, and God makes the world go round. I stop computing, the world still goes round. I shut my eyes wide and concentrate. I see a figure; it is a collapsing triangle. I am merely a node on these networked bio-quantum computers. I stop computing, they don't. I resonate, wouldn't you? But I still don't get it. I do not care; after all, this is junk, is it not? Should I pretend to understand? Not this time. I woke up by the railway, crossed the bridge, and went back home.

F **Rush and Burst**

It's time for another trip deep into the hyperreal, dear reader, to the Mecca of the hyperreal, in fact: Vegas!

I had been there once and hated it. I had stayed for a few hours only, waiting for my rental to reach the Grand Canyon. I did not care for the gambling and all the rest (I had already seen the Eiffel Tower and had a limited interest in a fake Luxor). But this time I had to come back, because of this paper I had found on the Web:

> This paper proposes a "jumping DNA" or transposon-mediated mechanism to explain rapid and large-scale cellular changes associated with human bodily transformation . . . A 1700 base pair DNA sequence was isolated from purified activated human T cells . . . The sequence of this DNA contains a novel combination of all three transposon families (SINEs, LINEs and HERVs) arranged like "beads on a string." I describe its structure and I propose that this DNA sequence, because of its cassette like configuration and its transcriptional expression and regulation, would be an effective participant in large scale transposon mediated genetic change that eventually results in transformation of the human body. The hypothesis is testable by using the DNA sequence as a molecular probe to monitor transposon activity in the blood cells of individuals undergoing profound psychological transformation as a result of advanced meditation, near death experience (NDE) or close encounter experiences with UFOs. The relevance of these proposed experiments to the study of survival of human consciousness after death is discussed.[47]

I can still remember my excitement the first time I read these lines. I passed them over to a friend of mine who is a science-fiction writer, and he diligently made it the centerpiece of the novel he was working on at the time. All kinds of junk represented on one short sequence, links with NDE and close encounters, all kinds of junk indeed . . . And that last sentence, almost matter-of-factly . . . and, by the way, there is this question, you know, of the afterlife . . . That was definitely too good to be true. It had to be a hoax, a prank of some sort. So I followed the links, did my little online routine, my hyperreal investigation. I could not believe what I found either.

The author, Colm Kelleher, whom we already met in chapter 1, was real, with credentials and all. All I found led me to consider him as a legit student of science for most of his young career; good diplomas, postdoc, and beginnings of a research career in good laboratories, with the expected mobility of a foot soldier of mainstream science; a reasonable list of publications, in reasonable journals. And then suddenly, a drastic change happened. Here he was, working for this private research institute lavishly funded by a Vegas billionaire,

Robert Bigelow's National Institute for Discovery Science (NIDS). The tone of the Web sources changed, and so did the vocabulary: suddenly, it was a question of anomalies, UFOs, Bigfoots, poltergeists, and the like. Some rumors on the Web were talking of NIDS as a CIA front. It was the *X-Files*, fair and straight. I was left wondering if he was a masculine version of Dana Scully, or a real-life Fox Mulder; or maybe both in one. And in Vegas, no less! I sent him an e-mail. He answered. I asked for an interview. He agreed. There was nothing weird in the exchange. I felt like I would have for any other interview with any other scientist. And yet, I was going to Vegas, for a crash course in anomalies, *X-Files* style.

I did not care about Bigfoots, poltergeists, anomalies, and close encounters per se. I was agnostic about that. They might have existed; they were not my problem. Already once in the past I had had the same experience: for my first book, I had interviewed one of the most famous UFOlogists on earth—about something else. And I had blown it. He happened to be also a successful venture capitalist. We talked over a great lunch in a nice restaurant in downtown San Francisco. We did not exchange a word on the questions that had made him famous. His name is Jacques Vallée, and he is a charming man, a great man, one who had the honor to be portrayed on-screen by François Truffaut. His story gave flesh to the last chapter of my book. During the editing process, my "development editor," surely a bit too aware of the codes of conduct of a university press of renown, insisted that we introduce him as "Jacques F. Vallée, who after his ARC career [Doug Engelbart's laboratory at Stanford Research Institute] became a novelist and a student of unidentified flying objects (UFOs) and of putative alien contacts with earth dwellers."[48] This might have been the only sentence I did not write at all, but I sheepishly let it slide, to my shame. The almighty ironical Metatron now grants me the opportunity to make amends: Jacques Vallée is the only recurring character in my two books. Jacques Vallée, like Edgar Mitchell of IONS, the sixth man on the moon, was a member of the NIDS scientific board.

So I was not exactly new to the process, although the process was slightly different this time around: Kelleher's paper implied that there could be some sort of connection between junk DNA and *some* anomalies. I wanted crackpots, kooks, and cranks, and it looked like I had my wish.

Quite naturally, this alleged connection linked to the most mysterious part of junk DNA, the part that most escaped the rigors of the standard model of molecular biology: the mobile or transposable elements (TEs). TEs, indeed, challenge greatly the ideal picture of DNA as a Turing machine provided by the standard model: how could "dead memory" ever be mobile? There is no review of the genome organization and evolution nowadays that does not stress the fluidity of the genome due to TEs' action, and the end of the "beads on a string" model. Barbara McClintock's work has finally been recognized. It took a while, though, since she discovered mobile elements in the maize genome in the mid-1940s![49] For a long time, she could not understand why her work was so "heretical," except maybe for the "tacit assumptions" most of her colleagues were promptly developing on the successes of the then-emerging field of molecular biology: "they didn't know they were bound to a model, and you couldn't show them . . . even if you made an effort," she told her biographer.[50] In different words, she had fallen victim to her colleagues' *trained incapacity.*[51] It took the "independent discovery" of the same kind of phenomena in bacteria, in the early 1970s, to realize that they were actually rediscovering transposition.[52] She was eventually awarded the Nobel Prize in medicine in 1983, "for her discovery of mobile genetic elements." Here are the concluding statements of her Nobel lecture:

> In addition to modifying gene action, *these elements can restructure the genome at various levels, from small changes involving a few nucleotides, to gross modifications involving large segments of chromosomes,* such as duplications, deficiencies, inversions, and other more complex reorganizations . . . In the future attention undoubtedly will be centered on the genome, and with greater appreciation of its significance as a highly sensitive organ of the cell, monitoring genomic activities and correcting common errors, sensing the unusual and unexpected events, and responding to them, often by restructuring the genome.[53]

The future has proved to be up to McClintock's anticipations, to a certain extent at least. The Human Genome Sequencing project has estimated that at least 46 percent of the human genome is made of transposable elements (TEs), "but [it] may yet prove to be much higher when diverged and degraded TEs are fully included."[54] Numerous voices inside the standard paradigm now insist that TEs might be the most important engine of change in evolution, and "looking forward, it seems likely that lateral DNA transfer will be increasingly in the news."[55]

However, even more molecular biologists still consider that TEs confirm the selfish DNA hypothesis of Crick et al.[56] TEs are thus mostly considered as molecular parasites (it was Britten's word of choice to characterize junk DNA too), even if another, and more reasonable, hypothesis is already available:

> TEs are often summarily dismissed in the literature as simply being selfish or even junk. It has been argued more recently that a more accurate and enlightened approach is to consider them and their hosts in the coevolutionary terms

> of host–parasite relationship . . . Such an approach is considerably more flexible, and envisions a continuum from total parasitism (or selfishness) at one extreme, through a middle ground of neutrality, to "molecular domestication" (or mutualism) at the other extreme. Indeed, the relationship between an element and its host may vary along this continuum over time.[57]

Some even consider that such types of relationships are essential to build a renewed conception of evolution, "a twenty-first-century view of evolution" that would lead to "to novel computing paradigms that may prove far more powerful than the Turing machine–based digital concepts we now use."[58] James A. Shapiro, who put forth this proposition, even had refueled the Neo-Darwinist versus intelligent design controversy when he claimed, some years earlier, that there could be "a third way" out of their "dialogue of the deaf":

> Our current knowledge of genetic change is fundamentally at variance with neo-Darwinist postulates. We have progressed from the Constant Genome, subject only to random, localized changes at a more or less constant mutation rate, to the Fluid Genome, subject to episodic, massive and non-random reorganizations capable of producing new functional architectures. Inevitably, such a profound advance in awareness of genetic capabilities will dramatically alter our understanding of the evolutionary process.[59]

But there were even more revolutionary ideas in Barbara McClintock's mind. In the closing pages of her biography, Evelyn Fox Keller spends some pages justifying her title, *A Feeling for the Organism*. She describes McClintock as somebody who "took pride in being different," a self-proclaimed "mystic." She tells about her interests in extrasensory perception, hypnosis, and various practices of biofeedback. She recalls her fascination for some Eastern practices, and especially for two "kinds of Tibetan expertise": their abilities to "run for hours on end without any signs of fatigue" and to "regulate their body temperature."[60] She even quotes her saying that "things are much more marvelous than the scientific method allows us to conceive . . . so you work with so-called scientific methods to put it into their frames *after* you know. Well, [the question is] *how* you know it. I had this idea that the Tibetans understood this *how* you know."[61]

Thus it might be some kind of poetic justice that led a real-life *X-filer* from Vegas to follow in both of McClintock's footsteps, on the exo- *and* the esoteric paths . . .

Here again, dear reader, I reach the end of my analytic leash. No more comments from my part, then, but six excerpts of the edited transcript of my interview with Colm Kelleher follow. You will be the judge, if ever you are in a judging mood (I am not).

Vegas's Real-Life Fox Mulder

TB: People don't look at you as if you were straight out of *The X-Files*?

CK: Friends of mine do, outside this organization, but not media, no. No. But yes, there are definitely parallels, we are doing almost precisely what some of the episodes of *The X-Files* portray. They touched on stuff that is very accurate. And actually, I was surprised at times. Well, I was impressed, some of the times, when, say, Scully would go and run a test for finding out the identity of somebody, they got the gels right, they got all the methodology right, so it is not kind of slapped hash. Somebody was working seriously to advise them on that.

TB: Do you look at that as a danger, because those references in popular culture could make people look at your work this way? Maybe people would take you less seriously because there are all those references in the popular culture?

CK: Possibly. I don't know what my next step is going to be after this, so it will all depend on that. I know that in general, all the people in our science advisory board are not affected. I know for a fact that they are not affected by being associated with this organization. So they are not affected or infected by this whole sort of popular culture part of it. Jacques Vallée has a very successful career in the venture capital business, and so meanwhile he has published ten best-selling books in the UFO field and people come to him at conferences, they are usually high-profile people who don't want to talk publicly. I've had the same experience. I have been at conferences and top-level NASA officers had come up and said, "You know, you've got a big fan base here at NASA headquarters, but nobody admits it publicly, but everybody is a big fan of that kind of thing."

TB: There are some rumors on the Web that you have big fans at the CIA too. Are you aware of that?

CK: Well, I know we have been accused of being a CIA front. Is that what you are talking about?

TB: Yes.

CK: Yes. It's just a natural thing because we are slow to publish, and again it's because the science advisory board hold their feet to the fire, and we publish some stuff, very little, of what we do; most of the stuff that we do is not published. Therefore, secrecy . . . These guys are obviously a government front and several members of our science advisory board have pretty high-level security clearances with intelligence agencies, so they lend this aura of spookiness to the whole thing, so there is a real *X-Files* aura with the science advisory board.

The Truth Is Out There

TB: You have been here for seven years now. Do you feel some changes when you talk about what you are doing, for example? Do people listen more now than six or seven years ago?

CK: Yes. I think that's probably accurate. I think the media is slowly sort of moving in the direction. And you have to say this is partly driven by ratings. There is more of an acceptance. Certainly in the media there is less this kind of little green man kind of thing than there used to be. But mainstream science, I think there is a lot of closet people, there is a lot of people in science who are closeted. Very much attracted or very much drawn to aspects of these kind of concepts that would never admit it publicly because their jobs would be in danger, the tenure track . . . which is understandable, but there are very strong cultural barriers to sort of letting all this stuff out.

TB: You don't even need to talk about extraterrestrial or close encounters or stuff like that. You told me at the beginning that when Crick published this paper in *Nature*, it basically killed grants on noncoding DNA. That is not an extraterrestrial thing we are talking here, we are talking about noncoding DNA, and already that seems to be taboo for some people.

CK: I know all of that kind of thing has dramatically changed. If you look, when I was doing a lot of the work in the late '80s early '90s on noncoding DNA, it was impossible to get money to study this stuff. And the person I was working under in Vancouver, Dixie Mager, she was writing grants and not much was happening. But now it's just become much more respectable. And now the sequencing of the human genome has become extremely respectable. It's getting more and more so as time goes on because there is a lot of evidences that are emerging now, introns play a role in regulation gene expression, there is Barbara McClintock's work in terms of the way transposition and that kind of thing can shape genome evolution. That's becoming more accepted.

TB: There seems to be kind of a hierarchy. In my mind, the most mainstream of all this is trying to look at some regulation phenomenon in noncoding, a lot of people seem to be doing that and that seems okay, right?

CK: Right.

TB: Then there is transposons and stuff like that, it seems a little less okay but still okay. And then there are these guys like Fritz-Albert Popp, the guys who are looking at light emission.

CK: There is a continuum there.

TB: Yes. Starting with biophotons, Popp told me, for example, he has been doing it for more than twenty years now, he told me even ten years ago people looked at him as a charlatan, as a crazy guy. Then there is the kind of things that you guys do, which is even less accepted.

CK: Yes, definitely.

Transposition Burst

TB: What could be the cause of what you call a transposition burst? Could it be some message coming from outside, or something from the inside? Would the antenna be a metaphor to talk about that?

CK: It could be, but the bottom line for me would be: the transposition burst, it would be a rare event, number one. But it would not be coming from outside. It would come from inside. So, suppose the elements of what Jeremy Narby would be talking about. It's also kind of a long tradition in the esoteric traditions that really the action is happening inside, it is not really coming from outside. And that's one of the reasons why I find it difficult to relate to this whole extraterrestrial thing.

TB: Although in your paper you make the point that near-death experiences and close encounters have the same kind of physiological effects or something like that so . . .

CK: Yes, there may be outer signals in the environment, but the primer, the driver for DNA change is coming from inside. You could see them as provoking the basic mechanism, but the basic mechanism is from inside. The whole idea is over a lifetime, over many, many, many years, and even decades, there is a slow process that happens. And it is intentionally induced. And it culminates, or it can culminate, in a transposition burst. And this transposition burst can completely reorganize the body, it is the equivalent of a metamorphosis. For a long, long time obviously I've been interested in that borderline between molecular biology and quote unquote sort of spiritual tradition. What are the consequences, the biological consequences, over a long time period of this kind of thing—independent, like I say in the paper, it is independent of the culture, it's independent of the beliefs, really, it's more like a basic biology . . . We are not talking about esoteric thought process, we are talking about a deep, ingrained physiological response, which, of course, has biological consequences at the level of the organ, and even below that at the level of the cell. So that's where my interest originated. The second part of that, of the whole thing, is that I did have an opportunity many years ago to take blood samples from a couple of people—you cannot call it a controlled experiment because it was RNA from a person who had been really sort of spending decades in this kind of mode, versus a couple of others we kind of grabbed off the street, so to speak, and there was a strong signal at the RNA level in this one person. However, we are talking anecdotes here.

TB: What about light? It seems to play an important part in both kinds of experiences, right? Enlightment, light body, but also every time people talk about close encounters there is immense light and . . .

CK: Yes, right across the board! There have been several books written on that sort of correlation between near-death experience, going down the tunnel of light, versus this kind of tremendous amount of light associated with close encounter experiences. The abduction phenomenon is like blinding light either initiates it or is part of it. So, you can relate that to the whole shamanic experience of going into the light, and going into a different world, and there are a lot of parallels there in terms of personal experience and the mystic, the so-called mystic quest. So, I would see this whole thing of, just different descriptions of light coming from different angles, as being fundamental to this whole thing. People have talked about this at the moment of death, these various light and beings, whatever description you want to come out with of light, the so-called light body. People when they are in deep meditation, and in various altered states also, see the same thing, a light body emerging from people, a light body merging with people, different sort of humanoid type of life-forms; even when I am studying anomalies, people call up and talk about that kind of thing. So, yes, that would be a very central part of this whole thing . . .

TB: In your paper you said that transposition burst could have effect on muscles, bone, but you said also hair and nails, they are the same.

CK: Carotene, yes. The reason that I focused on myosin and actin and tropomyosin and carotene is because they are the kind of fundamental structure. And one of the ways that you could visualize the transposition burst, and I am talking nonrandom transposition here, would be in effect the complete downgrading of the physical human being. And simultaneously, the activation of another set of genes. So, you've got down regulation, you've got up regulation, and you've got the emergence of a completely different, transformed body. That is semiphysical. There are a lot of legends in the multiple-personality disorder literature where people talk about eye color changes, people talk about things like tremendous allergy. Different personalities having different strong allergy responses. So, we are talking basic level of immune system, we are talking about gene activation on off on off depending on which personality. So, there is nothing inherently magical about this whole process; it's a fundamental biological mechanism and it just happens culturally that we have moved a long, long distance away from accessing it.

Noncoding DNA

TB: Back to this junk DNA, just stepping a bit outside of your published work, just trying to make a larger picture. First, how close are we, do you think, to getting some kind of an idea of what is in there? If there is something there, in noncoding DNA?

CK: I think it's only the beginning. What I said was relative because I think 99 percent of molecular biology and proteomics, and all of the spin-offs from molecular biology, are focused exclusively on 2 to 3 percent. You do have that 1 percent, which is bigger than the 0.1 percent of ten years ago. But we are still talking about a very small minority of people who are interested in noncoding DNA. I think we are only at the beginning. The idea that 97 percent of DNA is worthless or doesn't have any function is still a tendency in molecular biology to be pretty dogmatic about it; the whole biotechnology, for example, is focused on the 3 percent. The money, the cures, etc. However, at the same time, it's getting more and more play in molecular biology. The huge amount of this noncoding DNA is transcribed. Okay, so you've got a different level of information. People are getting very in tune with the RNA level in the last maybe five years since, especially since RNAi came out, which is the inhibitory effect of RNA. That is really sort of the beginning. People are beginning to see that RNA is fundamental. So, I see that as a very positive thing because it's very obvious that once things are transcribed, they have a much greater impact on the cell. And there is a huge amount of noncoding DNA that is transcribed. And in fact, the transition from the original DNA sequencing effort, when they went up one level and they looked at the cDNAs there was a dramatic, there was a ten to one, practically a ten to one ratio between the amount of cDNA that they were sequencing versus the genomic DNA, and all of that was transcribed, noncoding DNA. So, I see that as the beginning of a very big research drive probably in the coming decade where this stuff is going to become obvious that the RNA level of the cell is very important. Now, that has implications because you are probably aware of the RNA world that existed according to Thomas Cech, existed. I would speculate one of the offshoots of this nonrandom transposition burst, the emergence of the light body, is that this may be purely RNA-based as opposed to proteins-based. And intelligent, autonomous, you could call it an organism you could call it whatever, but at the level of RNA, so I think that the study of the RNA and the realizations of the importance of RNA is a very good start along this whole pathway.

Retro Code

TB: You said also in your paper that there was this controversy about the potential relationship between retroviruses and retrotransposons. Is this less controversial now than it was when you published?

CK: There is a whole chicken-and-egg thing there with retroviruses and retrotransposons. The overwhelming majority of people would say that retrotransposons are relics of retroviruses. Howard Temin was more in line with the idea that retrotransposons may give rides to retroviruses, chicken and egg, so I have to say that from an evolutionary point of view and from a DNA evolutionary point of view retrotransposons giving rides to retroviruses, to me it's a very credible way of looking at it. You know, Temin's hypotheses were very controversial.

TB: Would it be totally stupid to think about DNA as an ecology, not as a fixed memory? You know, the way we were taught at school, DNA is like a Turing machine, it's an infinite strand and there are boxes in there and they don't move, etc. The last twenty years, we discovered that boxes move.

CK: Well, it's Barbara McClintock's work, really, which dates fifty years, I would say. She was working in the forties and the fifties. So. Yes.

TB: But since 1977 we know that there are introns. In 1983 she received the Nobel Prize. Now some people say that retroviruses and retrotransposons may be connected. We see some cases of diseases, some very interesting diseases like mad cow disease and stuff like that, moving from one species to another. Therefore, could we think about DNA as an ecology, not some fixed form of memory, something that is opened, that changes in ways we didn't think about when molecular biology was created?

CK: I think so, definitely. I would say, and I think again Barbara McClintock was instrumental in that whole thing. The title of her Nobel Prize winning acceptance speech was "The [Significance of the] Responses of the Genome to Challenge." The genome was seen as an ecology practically. That constantly there was a dynamic shifting, changing, subset of thousands of different components. Yes, constantly changing. I would see that as being more and more, I think more and more people are seeing that and it is becoming more mainstream all the time . . . One of the interesting things about retroviruses is, during the life cycle, when they are transcribed and then they are translated, they co-package two different retroviruses that are completely different and then recombine in the subset and then you've got potentially bits of genes that are being swapped, but retrotransposons can co-package with retroviruses too and you've got the same thing, you could actually build a retrovirus from serial passages of retroviruses and retrotransposons.

Pffff!

CK: Most of what I have said today, if you played back the tape to a mainstream molecular biologist, he would say pff.

I was ready for a trip to Death Valley. On my way there, I passed by Area 51, and waved.

Part II 5' Molar Junk: Hyperviral Culture

But in this world that I'll leave, damned your eyes, to pay my dues only, and where I am already jigged, caught in the mousetrap of all the good reasons to live, men, dear men, I offer you with great pleasure,

1—My human condition,

2—I lay on the ground in this world where one is better off as a dog . . .
Here I feel in myself as a man in a house built while he was away . . .

—GASTON MIRON, *L'homme rapaillé* (my translation)

Chapter 4, All Tail

Close Encounters of the Fourth Kind

> Welcome my son, welcome to the machine.
> Where have you been?
> It's alright we know where you've been.
> You've been in the pipeline, filling in time,
> provided with toys and "scouting for boys."
>
> —PINK FLOYD, "Welcome to the Machine," 1975

Z The Crime of the Millennium

4.22.2002. In my wife's hospital room, the morning after the birth of our son. On the cover page of Montreal newspaper *Le Devoir* I read three headlines (here translated): *Genetic Therapy: The Lack of Funding Is Slowing Down Research, Is the Universe Digital?,* and for the Monday Interview section, *We Ought to Forbid the Patentability of the Human Being: We Are Not a Source of Commodities, Claims Maureen McTeer.* Late modernity enters the next millennium: welcome to the next level, echo Sega™ and Heaven's Gate™. Later in the interview column, I read about the recent decision of Canada's Health Research Institutes (IRSC) to authorize research on embryos before the project of the law aiming at regulating these activities is even discussed in parliament. Mrs. McTeer is upset. She thinks that it is not "a democratic way to proceed." She seems surprised, and unhappily so, that IRSC (which she calls "a lobby, which uses public funds, with a vested interest in such a high public priority question") can anticipate the law, and, more important, public opinion. Yes, how can scientists dare anticipate popular consensus on such a high public priority, affecting, *dixit* Mrs. McTeer, "the rights of the person"? Hello, Mrs. McTeer, smell the coffee.

9.18.2002. When hacking the genome becomes a garage activity. A few months later, a friend of mine sends me a copy of a newspaper article by Yves Eudes from *Le Monde* titled "Les pirates du génome." While Mrs. McTeer was still lamenting over the official decisions of instituted science, some "rogue" scientists were giving her real reasons to despair, creating genetically modified organisms . . . in their garages. In the French article, I read this quote from Eric Engelhard, the rogue scientist interviewed, the master pirate of the genome, the maker of venomless bees: "Here in Davis, there is a strong community of ecology activists, and I don't know what they will make of my bees. I can also fear the reactions of the conservative Protestant churches, which are violently opposed to any kind of genetic engineering. But I too am ready to fight for my ideas. I believe in absolute scientific freedom and I will carry on with my project." I decided to investigate further. I found the confirmation of what I had expected in the following post on Wired.com:

> Eric Engelhard is bioengineering a honeybee. In his garage. He's part of a new generation of bioinformatics brainiacs—people improvising with computers and molecular biology—who are making it possible to move genomics out of the lab and into your spare room.[1]

Science ought to be free, like computing. I do Linux on recycled CPUs, why shouldn't I engineer organisms on the same machines? Aren't they universal, after all? Junk is looming on my mind, that day, as I read this, in awe. *Code Redux.*

12.26.2002. Rael and his goons announce the birth of the first human clone, Eve. A couple of weeks later, Rael himself confirms that he has stopped all testing on baby Eve, by fear of legal action from a Florida court.[2] Don't act. The media frenzy around the original announcement had made it pretty sure anyway that the performative power of the Logos would be at work here. Enters the false prophet, and I remember the laconic comment of the Bible of Jerusalem, Apoc. 13:18, "According to Saint Irénée, the number 666 symbolizes the totality of evil from the fall of the devil to the coming of the Antichrist. *His worshipers will therefore be human-numbers, lifeless mechanisms, victims of a permanent manipulation.*"

11.12.2003. The Crime against the Species Has Officially Begun.

One more debate in the Higher Chamber, and the French nation will have innovated the notion of crime against the species, thereby acknowledging that the genocide of Homo sapiens sapiens, the autogeddon, as Ballard calls it, has officially started. That same day, an ad in the December issue of *Wired* attracted my attention: it described a new toy, a DNA sequencer for kids ten years and older:

> My First DNA Sequence
> The same tech megatrends that are reshaping grown-up gadgets are revolutionizing kids' toys. Nowadays, youngsters can race nitro-powered remote control trucks, fiddle with programmable robots, and guest-star in the latest sitcoms. If

> those aren't sophisticated enough for your brainiac tykes, the Discovery Kids DNA Explorer helps junior scientists extract and map real deoxyribonucleic acid. As third-grade science projects go, this is light-years beyond the ol' baking soda volcano. Next step: cloning Fido.[3]

The genomics cottage industry is well on its way to shaping our future, and our kids will be ready for it. Ethics and law will follow, as usual, twenty years later. In the legislative process, the notion of "genetic identity" was repeatedly put forward. Now, what could this notion mean when more than 98 percent of the DNA bases still evade our understanding?

U **Post-Scripta**

> The legal committee of the United Nations' General Assembly voted on Friday (February 18, 2005) by a slim majority in favor of a non–legally binding agreement that asks member states to prohibit reproductive cloning and adopt legislation to respect "human dignity" and "human life." But the text, which one diplomat said was intentionally ambiguous, does not define when life begins. The final declaration asks member states to "prohibit all forms of human cloning inasmuch as they are incompatible with human dignity and the protection of human life." If adopted and approved by the General Assembly, the declaration is not legally binding, so there would be no penalties for countries that do not implement relevant legislation.[4]

As I read these lines, I am reminded of the image that made the strongest impression on me during the long night of the 2004 U.S. presidential election. In the wee hours of the morning on the East Coast, it was a picture from California. At that time, the election was still undecided, and yet here he was, full screen, radiant with joy, Arnold Schwarzenegger, the governor of the state. His whole state had voted Democrat, and still, here he was, overwhelmed with joy, in front of a huge wall of video screens where the dominating icon was the word *WON.* As with all elections, this one had served several purposes, including the referendum on multiple propositions. If Arnold was so happy, in spite of his state's voting for the opposing party, I guess it was because it had passed many propositions that would effectively change the future of his state, and, most important, the future of business in his state. The commentary insisted on one proposition among the many that were approved that night: a proposition to encourage and fund research and development on stem cells.[5]

So it seems that global capitalism has now entered its genetic phase, the phase of our encounters with machines of the fourth type. After the simple machines of the old societies of sovereignty, the motorized machines of the disciplinary societies, the information machines of the control societies, human beings now face—or will soon face—*genetic machines.* I follow here Gilles Deleuze's footsteps in his famous "Postscript on the Societies of Control," where he matches the first three types of machines with their corresponding types of society.[6] Deleuze insists that machines do not determine the becoming of societies, but rather "express those social forms able to give birth to them and to use them."[7]

In *A Thousand Plateaus,* Deleuze and Guattari already distinguished machinic enslavement and social subjection and related them to different machines: the former happens when "human beings themselves are constituent pieces of the machine that they compose among themselves and with other things (animal, tools), under the control and direction of higher unity"; the second occurs when "the higher unity constitutes the human being as a subject linked to an exterior object, which can be an animal, a tool, or even a machine."[8] The focus of this distinction is on the regulatory unit and its feedback on the human constituent/subject.

The human being is an alienated slave to the machine when the regulatory unit of the machine maintains him or her in the state of a component, expandable part of a higher unit; he or she is socially subjected to the machine when she reconfigures him or her as a subject. In this opposition lies the original alternative between "overcoding of already coded flows" and "organizing conjunctions of decoded flows as such" that Deleuze and Guattari attribute, respectively, to the imperial state/machine (first type) and the motorized machine of the modern nation/state (second type). Cybernetic machines, as machines of the third type, construct a generalized regime of subjection that aggregates machinic enslavement and social subjection as its extreme poles.

The latest episode in the modern civilization described by Deleuze and Guattari is the cybernetic decoding and organizing of the flows of human nature itself, DNAs and bits, to the point that one now feels compelled to complete their enumeration, be it "an animal, a tool, a machine . . . *or a human being.*" What about these machines, then, which reconfigure humans as both a subject and an "exterior object"? And which decoded flows are they trying to organize? Deleuze and Guattari say that it is what cybernetic machines do, and they are right. But there are cybernetic machines and there are genetic machines. While the former regulate components as such without being able to actually build them, the latter both regulate and build its components. The autopoëtic machine, or second-order cybernetic machine, is no mere motorized, regulated, cybernetic machine. It is no mere computer. It is tomorrow's biocomputer; it's an egg able to count.

In the same way that the prototype of the cybernetic machine of the third type, James Watt's governor, was born with the first fully functional motorized machine, the steam engine, genetic machines were born with the first fully functional computers, that is, personal distributed computing machines. Genetic machines differ from computers as the regulator differs from the steam engine: by one order of magnitude in a series of logical types.

(Re)genesis is taking over control. One might argue here that if it is the case, the "societies of control" will have been dominant for an instant only, especially when compared to the previous societies, that is, the societies of sovereignty and the disciplinary societies so well diagnosed and described by Michel Foucault. Several answers are possible here.

The first involves the argument of "the acceleration of history." Without even considering the hypothetical fourth type of society that I am advocating here, this phenomenon is quite obvious. When Deleuze writes about the societies of control and opposes them to the disciplinary societies, he gives various chronological references that could be interpreted as showing this acceleration: (1) Foucault locates the disciplinary societies in the eighteenth and nineteenth centuries, (2) the Napoleonic regime seems to mark the transition between societies of sovereignty and disciplinary societies, (3) the latter "reach their height at the outset of the twentieth century," and (4) they have "ceased to be" after World War II. So, whereas societies of sovereignty have lasted for centuries or even millennia, disciplinary societies had a life span of at the most a century and a half. Deleuze rightly credits Foucault for having already noticed the "brevity" of the latter.

A better answer might be, however, that these societies are not "stages" (à la Rostow),[9] but instead "becomings": each new type of society does not abolish the preceding type, but supersedes it. Deleuze even writes that "it is possible that old means, borrowed from the ancient societies of sovereignty, will return to the fore, but with the necessary adaptations."[10] In fact, it is crucial to understand that, as such, the establishing of new social forms (i.e., societies) does not belong to history: they are events in their becoming, and as such they are, as Nietzsche says, "untimely" *(intempestif)*. Deleuze is very clear about this point in an interview with Antonio Negri that composes the preceding chapter of *Pourparlers:*

> History retains of the event its effectuation in states of things, but the event in its becoming eludes history. History is not the experimentation, it is only the almost negative set of conditions that make possible the experimentation of something which eludes history . . . Becoming is no history; history only refers

> to the set of conditions, as recent they may be, that one turns away from to "become," that is to say, to create something new.[11]

In this perspective, societies of control might appear as a phase transition in the becoming of capitalism. I employ "phase" here not in the sense of "stage" ("a distinguishable part in a course, development, or cycle" says *Webster's*), but rather in its sense in the philosophy of Gilbert Simondon, as the *overtaking of the opposition between being and becoming:*

> One conceives of a phase only in relation to another or several other phases; there is in a system of phases a relation of equilibrium or reciprocal tensions; it is the actual system of all the phases considered together that is the complete reality, not each phase in itself, a phase is a phase only in relation to the other phases.[12]

Deleuze insists, and rightly so, that what is at stake here is the analysis of capitalism and its developments, that is, a political philosophy centered on the individuation(s) of capitalism.[13] Societies of sovereignty, disciplinary societies, societies of control, and, I argue, "genetic societies" are phase transitions in the becoming of capitalism. The actual reality of capitalism deals simultaneously with sovereignty, discipline, control, and soon enough with generation (of its *subjects*). It is not *only* a question of production (as in the disciplinary societies) or overproduction (as in the control societies, claims Deleuze),[14] but also a question of reproduction (or generation). "Machines," says Deleuze, "explain nothing; we ought to analyze the collective apparatus [les agencements collectifs] of which machines are a part."[15]

Rather than being mere anachronisms, several components of the collective apparatus of control described by Deleuze are in fact better understood, I claim, as transitions for the "next" phase of capitalism, that is, "genetic capitalism." This is the case of Burroughs's diagnosis of control itself, its peculiar way of turning human beings into addicts, "unnecessary" junkies (according to *the algebra of need*).[16] Burroughs was even more right on the money than he thought: junk *is*, indeed, "the mold of monopoly and possession."[17] This is also the case, according to the same logic, that "*Man is no longer man incarcerated, but man indebted,*"[18] and soon enough, man annexed as capital itself (see chapter 5), living money, the genitalia of capital (which is why this next phase of capital could also be called "genital capitalism").

Machines might explain nothing, but they are the engines of these phase transitions. It was once the world was untimely enfolded in a global network of personal (albeit "pumped-up") computers that the human being could be described as a genetic database (see chapter 2). The decrypted genome is the equivalent of the meter kept in the museum, an *etalon*. It is both metal and silicon, dollars and gold: a new universal equivalent. Gene banks are indeed the financial institutions of the machine/state of the fourth kind.

For twenty years at least, we have heard about nanomachines, artificial intelligences, artificial forms of life. For more than twenty years now some human beings have been busy building them. In the past fifty years we have described the structure of DNA, and decoded the genome base by base. Human beings, flies, mice, and some worms are now officially databased.

Here is my question: What becomes of ethics—if, as Foucault had it, ethics is the reflected practice of freedom—when we have already left behind the era of mass production of cadavers declared by Heidegger, and entered the era of mass production of genetic *goylemes*? First steps, baby steps in the slow process of commoditization of Man™ . . .

The process of genetically modifying a human being and growing it out of stem cells will, in all likelihood, be developed to scientific success in the next twenty-five years or so, what we used to call a generation.

Some groups, sects, and laboratories have already started talking about their attempts to clone a whole human being. A guy alone in his silicone garage has effectively done some species-changing genetic manipulations, and released into the wild the resulting monsters.

The French parliament has already invented the legal notion of a crime against the species, superseding the crime against humanity, and therefore acknowledging that the crime has already begun.

By the time that my son (or your daughter, or their sons and daughters, it doesn't matter) will reach his reproductive potential, there might be machines to produce superbabies (and, no doubt, underbabies).

In the meantime, today's kids will play with their brand-new DNA sequencers for children under ten years old.

So, what do you think about freedom now?

Freedo(o)m is still the name of the game. And nowadays too, experience as much as reason teaches us that men think they are free because they are conscious of their actions and ignorant of the causes that determine them (Spinoza, *Ethics III, prop 2. scolium*). Here in Quebec freedom is (and is not) a brand of yogurt. Hell, the whole world, down to its micropolitics, is pro-freedom (even yogurt manufacturers). Freedom of choice, pros and cons. Could you be against freedom?

And yet, do you feel free? Have you been liberated, lately? Or do you, like most late moderns, live in anguish before the magnificent power of the machine of the fourth kind, its haecceity slowly spelling: I can make you . . . I can make you . . . (this was Dr. Stern's diagnosis already in 1956). Do you feel obsolete? Expendable? (Please swallow your Prozac now if you feel depressed, and get back to work.)

Philosophy is no tranquilizer, it does not cure or alleviate: it is at best creative for both its author and its readers (I know they are supposed to be dead, but who isn't, these days?). What would be today's concept of freedom? How can we build it, in spite of the fact that it may soon be able to build us? My attempt at a starting point for an answer will be: search in junk! And, more accurately, start in our own junk DNA.

What we now understand of DNA through the so-called genetic-code metaphor is the message coding for the human survival machine, the material body. Orthodox neo-Darwinism claims that it is all there is to it. Actually, in the most obvious nominalist fashion, it has *ruled* (for a while, as long as rules last) that it is all there is to it. DNA coding for protein synthesis represents approximately 1 to 3 percent of the whole DNA. Junk DNA is a common name for the rest, more or less. What if, indeed, the quest for the Body without Organs should start, quite literally, in our own junk?

V **(Bio)Ethics**

By the time these few words will be put in print, it will be well over ten years that the human entity usually called Gilles Deleuze has been dead, and unfortunately for him, he did not put his faith in transcendence. So we are left with this open question: what remains of an ethics of immanence after his matter is gone, fed back to the pre-individual field?

We are left with questions. Ethical questions, because it was, after all, his trademark.[19] With Slavoj Žižek,[20] I agree only on this, no hyphen ethics, just ethics. There is no biogenetic ethical question per se: the ethical question remains the same albeit in new—and potentially extremely crucial—modalities. So today's question is still *what do you do with your freedom?* And Žižek is right indeed in raising the question of its modality: *how do these new conditions compel us to transform and reinvent the very notions of freedom, autonomy, and ethical responsibility?*[21]

The rest of Žižek's development, posturing so-called Catholic counterarguments to better dispel them, I am sorry to say, however, is just good for scrap, straw men and company: it leads, unfortunately, through Lacanian psychoanalysis, to the revelation that we were never free in the first place. Either Žižek has not heard of the Fall or he is quite happy to make it last. Indeed, he must be when he proposes to finish the Enlightenment project and "follow the logic of science to the end . . . waging that a new figure of freedom will emerge."[22]

Reading these lines, I was reminded of the end of the Appendix to *Foucault,* where Deleuze too makes the wager of "the advent of a new form," in relation to the same new modalities: the *superman,* neither human nor God, "which it is hoped, will not prove worse than its previous two forms."[23] There is hope in the overhuman, this form that stems from a new play of forces located outside of the human, in the revenge of silicon over carbon, of the genetic components over the organism, of the agrammaticalities over the signifier. Outside of the human?

In which ways did silicon supersede carbon? How did the genetic components supersede the organism? On their own? Did the sands suddenly express a new life force? No no no: man is still in charge, and overman is the compound form of forces in man with these new forces. Overman is the man *taking charge* of the animals, of the rocks (the inorganic life of silicon), of the being of language. Deleuze wrote, following Rimbaud,[24] *"l'homme chargé des animaux même (un code qui peut capturer d'autres codes)."*

Keith Ansell Pearson is right to point back to the magnificent formula of *Anti-Oedipus,* "man as the being who is in intimate contact with the profound life of all forms and all types of beings, who is responsible for even the stars and animal life . . . the eternal custodian of the machines of the universe."[25] But he follows the original English translation of *"chargé"* by "responsible." In *Anti-Oedipus* too, however, Deleuze and Guattari wrote *"chargé"* as if something or somebody (God, this previous form?) had loaded the human being with the stars and the animals,[26] had put man in charge of the machines of the universe as a "custodian" (*un préposé*). Man is held responsible for the earth, like a *préposé,* with the machines of the universe in his custody, a kind of super-Noah (or a free-floating ass).

All here is in the passive form, cryptic allusion to the rainbow of the covenant: "this is the token of the covenant which I make between Me and you and every living creature that is with you for perpetual generations: I have set my Bow in the cloud, and it shall be a token of a covenant between me and the earth" (Genesis 9:12–13). Note this: the covenant is with the earth, and all that is made of it, the living creatures, and man is its custodian, not for all eternity (as Deleuze and Guattari have it), but for *perpetuity.*

Note also that in this version of the story, man was never in charge of the rocks and stars, but only of what he gave names to (Genesis 2:19), of what was "delivered into his hands," more bluntly, of what he can eat (Genesis 9:2–3). Deleuze and Guattari thus extrapolated the original story, giving man charge of all the rocks and stars, whose sole custody was that of the angels, so far. No more need for angels; thanks to his technology, man has become a star eater, "has plugged an organ-machine into an energy machine, a tree into his body, a breast into his mouth, the sun into his asshole."[27]

In other words: when did man develop an appetite for the inorganic? When did man start consuming matter as such, not only living matter (albeit deprived of its running blood)? When did man start to watch over the celestial spheres? Forget about the "start," the origin, and let me rephrase the question in a better Deleuzian fashion: what about the starworm-becoming of man? Do you need to be schizophrenic to know the starworm in you? Do you feel the sunshine in your asshole?[28]

What do *you* think that you are made of anyway? Water, earth, wind, and fire? Stuff dreams are made of? Carbon, oxygen, hydrogen, nitrogen, salts, and metals? Star dust? From the biotic soup of a pre-individual magma, undifferentiated and monophased? The genes of your ancestors plus chance? *Hasard et nécessité?* Many voices talking in your head?

Are these mutually exclusive options? May I risk a synthesis?

Overman (transhuman, extropian, cyborg, ribopunk, name your brand): custodian of the machines of the four kinds, hybrid carbon/silicon life-form of the future, organizing flawlessly the conjunctions of decoded singularities (Deleuze and Guattari), group individual twice dephased and open to the multitudes of his *milieux* (Simondon). Overman, master of DNA, breeder of men (Sloterdijk, after Nietzsche and Heidegger). Overman, the next phase of the becoming-starworm of man.

Superman, the next proper name of the autogeddon, equipped with the best logic science can provide. To the end! Let us get into abstract sex,[29] let us go capture other codes . . . Let there be monsters and chimeras, parthenogenic babies and clones. Let the better over(wo)man win![30]

G **The End of a Common Nature**

Žižek also wrote:

> The main consequence of the scientific breakthrough of biogenetics is the end of nature. Once we know the rules of its construction, natural organisms are transformed into objects amenable to manipulation. Nature, human and inhuman, is thus "desubstantialized," deprived of its impenetrable density, of what Heidegger called "earth."[31]

This fiction holds all in its "once"; it is a false premise today and still will be for a while. All that this alleged biogenetic breakthrough does is make you believe that man indeed knows the rule of the construction of *a* natural organism. Or could know it soon. Yes, man has started to decode the fluxions and superfolds of DNA. But today this knowledge is still rudimentary, mechanistic, and dogmatically (and centrally so) oriented.

Let me summarize and complete the picture: there is junk in life.

First man ate earth and re-created earth;[32] it was the becoming-earthworm of man. Then man plugged an organ machine into an energy machine, and started watching over the stars and sands, building machines to reach them, and be touched by their profound life. Soon earth was covered with machines and sands were clocked in megahertz. Man started to eat machines and created other machines, organ machines, engines and batteries. Man plugged the organ machine and the energy machine into a metamachine, an information machine. Man plugged all the information machines together and started a decoding feast upon himself: man started eating man and creating overman. Unfortunately, overman soon realized that eating man was the wrong name for incorporating humanity: it only consumed the material body, one at a time, even if by greater numbers as the centuries passed. The earthworm was left with ashes, longing to eat the stars and his mouth full of junk: is it in the ether where a worm should live?

In the ether of number?

Here is a reminder: At the beginning of the fourteenth century, this is what Duns Scotus considered the ineluctable consequence of nominalism: *If all real difference is a difference in number, the difference between genus and species is a mere distinction of reason.* Therefore, an individual could not generate another individual of the same nature, and generation and the unity and fixity of natural species would be a concept without foundation in the mind.[33]

If all real difference can be encoded and decoded in numbers in a difference engine, each singular becomes his own species according to number. Just like angels, following Aquinas, and in spite of his matter, this time. Just as data would decide a programmer. When the difference engine turned into the universal machine, difference in number became all there was to it: singulars living in discrete time, simulacra. If the computer is indeed the universal machine and man the custodian of the machines of the universe, soon the computer will remake man in its own image (and vice versa): it will be the end of generation as we know it. This could mean one end of nature all right, but not the end of all nature.

Here again Duns Scotus had a very convenient distinction to make sense of the present situation. Of which nature exactly would the consequences of the biogenetic breakthrough be the end? *A* nature and not Nature (be it *natura naturata* or *naturans*).

The God who made the world and all that is in it, the Lord of heaven and earth, does not dwell in sanctuaries made by human hands (Acts 17:25).

If, for Deleuze, Spinoza is the "prince of philosophers," Duns Scotus should be their king: he is the one who gives to being his only voice, this voice that makes *the clamor of being*, his univocity.[34] Deleuze insists; Duns Scotus represents the first of the three principal moments on the philosophical elaboration of the univocity of being:

> In the greatest book of pure ontology, the *Opus Oxoniense*, being is understood as univocal, but univocal being is understood as neutral, *neuter*, indifferent to the distinction between the finite and the infinite, the singular and the universal, the created and the uncreated. Scotus therefore deserves the name "subtle doctor" because he saw being on this side of the intersection between the universal and the singular. In order to neutralize the forces of analogy in judgment, he took the offensive and neutralized being itself in an abstract concept. That is why he only *thought* univocal being. Moreover, we can see the enemy he tried to escape in accordance with the requirements of Christianity: pantheism, into which he would have fallen if the common being were not neutral . . . With the second moment, Spinoza marks a considerable progress. Instead of understanding univocal being as neutral or indifferent, he makes it an object of pure affirmation. Univocal being becomes identical with unique, universal, and infinite substance: it is proposed as *Deus sive Natura*.[35]

In this strategic reading, Deleuze thus qualifies as "progress" the way Spinoza fell into the trap that Duns Scotus wanted to avoid: pantheism. Clearly, for Deleuze, pantheism was no trap in the first place, it was merely a (dogmatic) "exigency" of Christianity to avoid it. When he argues further that Spinoza's progress is incomplete, he makes it clear that "such a condition can be satisfied only at the price of a more general categorical reversal according to which being is said of becoming . . . Nietzsche meant nothing more than this with the eternal return."[36]

Here are the three moments of the elaboration of the univocity of being according to Deleuze: Duns Scotus and the first, "indifferent," ontology, Spinoza and the first dispelling of indifference into "an object of pure affirmation," and finally Nietzsche with the eternal return that "is the univocity of being, the effective realization of this univocity."[37] From Duns Scotus's indifferent being to pantheism in Spinoza to the death of God in Nietzsche: "in other words, creation, 'being' which was understood by nominalism as the radical dependence of all things on the will of God, now becomes (in valuation, in the concern with the Nothing) the radical dependence of all things (for their meaning, their existence) on *me* as subject."[38] This is a necessary digression, but this is not my point. I'll stick with Duns Scotus here.[39] Because it is not only with the univocity of being that Duns Scotus is useful here, but also with his notion of individuation.

For Scotus, individuality is what is such that it can only be in the subject in which it is, that is, in this-subject; a nature is what is such that it can be in this-subject and that-subject. A nature is thus a common nature, individuable and nonpredicable. The concept of this nature, like all concepts, is predicable and nonindividuable.[40]

A nature is not reducible to its concept: it exists out of the mind (as Philip K. Dick used to say, reality is what is still there when you stop believing in it). Duns Scotus's ontology (like Deleuze's) is realist.

So, which nature is going to end—according to Žižek and many others—with the advent of the machines of the fourth kind (i.e., the biogenetics breakthrough)? *The common nature born out of sexual generation.* (As we've known it so far.)

The question *Is this-subject genuinely alive?* could become an actual ethical question for some. Some others will enjoy the opportunity to decide that this-subject, not being genuinely alive or, later, being genuinely alive, should be considered as an underman. And the same war will return again (between the ancients and the moderns). Up to the next level of the eternal return!

Or then again, maybe not: there is junk. Does junk DNA entail another common nature, of a kind not yet imagined? Some hypotheses beyond the selfish doctrine already exist. Some argue that crucial regulatory functions are buried in junk DNA; others have more challenging theses in the works. Today's truth is that nobody knows what junk DNA can still hide.

For Junk DNA is the black matter of the ontogenesis of the machine of the fourth kind.

The dunce concept that I have in my guts is that DNA is a whole unity, and not only by numbers, continuous or discontinuous quantity, matter or bare existence. That the pre-individual could after all be in each individual; that each singular DNA participates in a whole DNA ecology. And all living beings are connected by the powers of junk.

T **Individuation, without a Principle**

7.25.2006. There never was an individuation *principle.* Many searched for one, though, and for a long time. Some felt that they had found one: in matter, form, number or quantity, space or else. The brightest, the subtlest, ended up with some twisted tautologies. *Hic et nunc,* I know one when I see one. Some argued against nominalism but resorted to a lexical innovation when cornered. Eventually, at the sunset of modernity, some others concluded that the question was surely badly put if there were still no answers to it after centuries of effort. Hence, the inversion: there is no individuation principle, because . . . there are no individuals. Back to square one, that is, pre-Socratics (or, more precisely, ante-Aristotelians). Not twice the same river, no identity principle valid in this quest, and no sense for an excluded third. Instead: a process, a never-ending process; no individuals, only individuations, and time comes back with a revenge that some call *eternal return.*

Gilles Deleuze, after Whitehead, Bergson, and Simondon, synthesized this process philosophy. No Being but one being, relentless becoming. Univocity of being was not a choice, but the only logical solution, the only way to hear the "clamor of being." Then Desire expressed itself, down to the eternal return spiral. A dangerous spiral at that: there is no comfort in spirals, confirmed Trent Reznor, whose nails are actually longer than those of the Master himself. The truth is, we do not need an individuation principle, and *principle* is usually the name we give to our desperate attempts to bring back some order in the equation. Order is but an exception: always fragile, doomed. So, instead of stability, one started to talk about *meta*stability (Simondon) and dissipative structures (Atlan, Prigogine). All right, then what? Ah, the comforting feeling of a lexical invention. OK: pockets of order, against all odds, against the dreaded entropy: chance and necessity. This pocket of order I call myself, I, this individual decision to recognize myself as such, an individual. Let me choose my becoming. For I do have this freedom, don't I? No, you might say it's just an illusion. Well, watch me (and eat it). Period. Why keep arguing?

Anyway, it's not my problem, and let me be bold and claim it ain't OUR problem no more. Individuals happen to be no more or less (un)certain today than they were yesterday (next, I will claim, only more dis-affected; see chapter 5). If one resorts to the *hic et nunc,* the Scotist solution, *haecceity:* they are still there, where else could they be? Today's most thorough investigation,[41] impeccably logical as it should be, led to the most obvious, it seems: what's an individual? Some noninstantiable being (yes, you can multiply a colon, but not an individual)? Haecceity is not instantiable, there can be one and only one instance of this particular *hic et nunc.* Suffice it to say, it works enough for me, but it is not my problem, it ain't OUR problem no more.

OUR problem, our urgent, extremely critical problem, is named: *common nature.* Or lack thereof. Yes, or lack thereof. Dreadful, isn't it?

At first, you resorted to the nice stories your parents had told you: baby boys are born in coleslaw, girls in roses; then came the charming metaphors, Daddy the gardener, with his little seeds, Mommy the garden . . . Then, once a teenager, you looked at the naked truth, first with revulsion, then with fascination (or vice versa). You had to just do it. Sex, that is. You started to notice it was everywhere. Tempting, alluring, dirty or not, with or without meaning, but *there.* Some chose to acknowledge it, some not. Some devoted themselves to it, some not. Some gave fancy names to it—Love, Desire—and some just did it. Some actually tried to reinvent it, some not. Everybody thought about it, it seemed. But whatever all this business amounted to, there was something you never doubted, whether you practiced it or not: this, for us humans, is where individuals are coming from. Rational animals maybe, but sexual beings for sure.

I was born in the early 1960s. I never knew a world where sex wasn't everywhere. It wasn't taboo anymore and almost anyone could answer the question "What's an individual?" with "The result of Daddy and Mommy making it."

Sure, animals do it too. But we humans *think* about it: what's a human being? Freud answered: some sex-obsessed creature, and developed a nice-sounding theory about triangles, with a charmingly tragic name. People started referring to *their* Oedipus . . . My very own Oedipus . . . Daddy, Mommy, I, and Oedipus, our silent partner in crime, in the crime that is MY life. Oedipus, the partner of my individuation and the measure of its relative success: this blind and limping character whom, knowingly or not, my parents created along with me. Oh, yes, he's real. Like a ghost floating over my cradle, like a personification of Daddy and Mommy's relation leading to me. I and Oedipus, Oedipus and I, we're best pals, again, unconscious partners in the murder of my father.

In the twentieth century, this came to be known as the unconscious tainted side of the individuation mirror. Physical entities may be individuated; living entities may be individuated as physical entities plus something else; and we humans may be individuated as living entities plus something else yet: Oedipus. Aristotle, his cultural contemporary, called him our "noetic soul," this something else that made us human apart from other living beings. Rational animal, that is, animal endowed with a noetic soul. Oedipus is the name of the archetypical (because individuated) noetic soul, tainted with desire. Oedipus is the stigmata of the third tier of our human individuation, the reflective feeling that there is something not so clear about the whole business of being human.

Oedipus is a name, hence a linguistic creature. There might have been an individual once named Oedipus, who was self-blinded, and limping. But most important, Oedipus came to be known as the protagonist of a human tragedy.

His tragedy became our lot, constitutive of our being human. Under this name, by some insight of one inquisitive noetic soul, Oedipus became a technology. In fact, Oedipus became the brand name of the late-modern technology of the self.

Simondon said that all individuation proceeds by transduction, propagation . . . *prise de forme.*[42] He picked up his physical paradigm with a lot of caution. The crystal. He argued from analogy up, with this paradigm (this instance) as a starting point. He moved to living entities as a second tier of individuation, and picked a coral colony as a second instance. Then, when it came to us, he added the last tier: psychic and collective, in other words, the noetic soul that makes us rational animals. A lot of other people pondered about that, silently, while he remained unknown. Then many came to the same conclusion at the same time: language and technology are the two distinct modes of expression of the noetic soul (its *grammatization*). Maybe, even, the noetic soul is language technology (it expresses itself self-sufficiently).

At approximately the same time, the latest revolution happened. Away with discipline! First: I do not need a Daddy and Mommy to censor me all my life. Drop out! No more policing my share of the noetic soul. Fuck Oedipus. Let there be freedom. Let me choose my parents, my friends, my community. I won't work anymore under some laws that were actually never convincingly explained to me. No more necessity? Pure chance? No, no, pure potential, chance *and* necessity. I don't care where I come from, I can still develop into something else, but something else I chose to be myself. Deliberately, in a definite gesture of self-expression, of self-production: *let there be me. Potential,* the latest big word of the latest revolution. A liberating metaphor: I am that grows. I do not need to ever be myself, just become myself. I am in control of my own becoming. Self-centered, self-sufficient, albeit altruistic and well-wishing revolutionary individual becoming, with many aspirations. Everybody's an artist when art becomes your very own life. Why not? Why not, indeed?

Simondon said, with a quite a dose of enlightened anticipation: *Nature is the reality of the possible.* Late-modern human beings took him at his word. In Duns Scotus's terms: this *haecceity,* this ultimate actuality, became, in Simondon's terms, the resolution of a metastable state, charged with potential. The individuation tautology had somehow become the quasi-tautological chiasm of modal singularity: common nature is virtual singularity, and the actual singularity is common nature in act. Or, as Paolo Virno puts it: the individual singularity adds something to the common nature, but without ever exceeding its perfection.

Duns Scotus's common nature had become Simondon's pre-individual, this virtual reservoir of singularity. Quite a successful quasi tautology indeed!

Virtuality, this Aristotelian intuition, came back to the rescue of the actual conundrum. If the actual was so decidedly aporetic, let the virtual be its tainted side, and so dispel the aporia, like any vulgar paradox. Everything that was a substance so became a process: from the Virtual to the Actual (actualization) and back, from the Actual to the Virtual (virtualization). Virtuality is pure potential, fields of potentialities. Some get actualized, others don't. Yes, but what about the possible? The possible is not the potential. Deleuze, after Bergson, concurred: the possible is only retrospective, it comes after the fact, as the name of this potential that did actualize (THE possible) and that could have been actualized (but was not, the possibles).

THE possible becomes Nature . . . by some process of individuation, by multiple processes of individuation. Because, as science has held since it has existed as *experimental* science, nature only knows individuals (bare, concrete individuals, particulars, singularities). The reality of the possible, nature, is and is only made of actual singularities, that is, of these possibles that individuated (these potentials actualized). All right.

But what about the old idea of a common nature, then? Could there still be such a *thing*? Nominalism would deny it as a mere concept (*une vue de l'esprit*), or worse, a name that would only describe something (nonexistent) by virtue of human convention. This nominalist perspective would insist that a common nature is what we agree it to be (no more, no less). Or, more classically put, "common natures are formally distinct aspects of entities that are the primary significate of a common noun." In this perspective, common natures are real in the sense that these formal distinctions are expressed in actuality, where they refer to some actual being; but they are not real in the sense that a common nature cannot be physically separated from these actual beings they refer to, and which only express the common nature imperfectly. In other words, a common nature can only be virtual (but virtuality is a mode of reality).

There used to be such a common nature of human beings: Daddy and Mommy must have done it. One could not conceive of a human being without him or her having been conceived first by the sexual encounter of his or her genitors. Crucially, in this old way, the common nature of human beings stemmed from the "natural" reproductive behavior of human beings, and this is why human beings were living entities sharing this with any other sexual forms of life (be it animal or vegetal). In other words, the third tier of individuation, the noetic soul, depended first on the second tier: the human animal was an animal first, and rational second.

But then sex became the main figure of irrationality. Oedipus showed his true face: that of a tyrant, ruling over sex without ever showing his monstrous face. You thought you were free, but you were actually governed by this faceless tyrant. This, Freud said, was the third (and maybe last) narcissist wound. No, not only you humans are not the center of the universe (Copernicus inflicted this one); no, not only you humans actually do not essentially differ from other forms of life, you are the mere results of evolution (Darwin inflicted that second one), but you humans are not conscious of what actually governs what you consider the most free of your acts. Alas.

These, after all, were the modern times, and modernity had found a way to deal with tyrants: behead them! Ah, sacred dialectics, which can make a servant out of a master! Let not Oedipus be your tyrant, recycle him into your servant: make a technology out of him. Make him a conscious actuality, and he will be your best friend. Oedipus is your parents' creature. He is not you, you can actually get rid of him and still be you, a freer you. Find your true self through the perfect technology of *oedipectomy* (also known as Freudian psychoanalysis). Domesticate your Oedipus, tame him or you will forever be his toy, his child victim (yes, Oedipus is a child molester). Seize control over your Oedipus. Yeah, more power to you.

Better yet: reclaim control over the Generalized Oedipus, because, you guessed it, everybody has his own little Oedipus. Oedipus is a social thing, a collective individual. Yes, everybody's his victim, even your Daddy (and, Melanie Klein added, even your Mommy), even your boss, even the cops—hell, even your psychoanalyst. Even more power to you if you understand this. But you do not need to be a victim forever: be your own Oedipus (instead of having an Oedipus complex). The revolutionaries said: liberated men are their own Oedipus, they are today's overman. Their model might have been schizophrenic and incensed for that reason, but, most important, he self-proclaimed: "I, Antonin Artaud, I am my son, my father, my mother and myself; leveler of the imbecile trip where begetting tangles itself up."[43]

How else do you think that the last revolution could be invented, other than over the ashes of a potent mix of Freudianism and Marxism? Fortunately, when this revolution was brewing, science came to the rescue, and at last found a perfect candidate for some sense of continuity. Yes, there was a new basis for heredity, a new characterization of forms of life that could restore some sense of belonging to the realm of the living. Let there be DNA, this material universal. No more need for Oedipus, then, DNA will do:

> The unconscious is not fundamentally a repository of submerged feelings and images as in the vulgar Freudian model. Neither is it fundamentally a Lacanian dialectic between the Imaginary . . . and the Symbolic . . . It can be made to be

these things, on two of its levels. More broadly, though, the unconscious is everything that is left behind in a contraction of selections or sensation that moves from one level of organization to another: *It is the structuration and selections of nature as contracted into human DNA.*[44]

To be honest, I must confess that I truncated the quote here. What follows is a series of "levels," "syntheses," and "thresholds" that articulate the specific form of becoming that the unconscious is. I forgot the rest of the series because I want to stress the first "level," the "molecular stratum," and note that, in agreement with their time, Deleuze and Guattari found this level in DNA. Following François Jacob and Jacques Monod,[45] they offered that the unconscious syntheses, at this molecular level described by "microscopic cybernetics," have no regard for the "traditional opposition between mechanism and vitalism."[46] Only desiring machines and molecular unconscious, no more castration and Oedipus: here is Deleuze and Guattari's new *mot d'ordre.* But it is a dangerous *mot d'ordre,* especially if you equate the "molecular" with the "genic," DNA with the genome.

The central assumption is that molecular biology has "proved" that DNA is the medium of a universal code characteristic of all forms of life. Mitosis and meiosis, the two basic operations of, respectively, asexual and sexual reproduction, were shown to be processes involving DNA. Like all forms of life, humans were understood as DNA-based. The old processes of embryogenesis and development were (and are still) translated into subtle hypotheses about the mechanisms of gene expression and regulation. In the early 1960s, Francis Crick and Sydney Brenner held, no doubt, that molecular biology would soon have solved the "last two problems remaining in biology": development and consciousness. Simondon's second tier of individuation had found a new basis. Sex as we knew it was recast into *one* strategy to perpetuate DNA, along with asexual reproduction, which soon became the post-Oedipian phantasm of choice for late-modern individuals.[47]

In this new scheme of things, DNA became equated to a—maybe even *the*—common nature: all forms of life have DNA; DNA defines life, and the genetic code is universal to life. On earth, life is carbon-based, and atoms of carbon are the building blocks of proteins. Proteins are the stuff of life, and DNA is the blueprint for protein synthesis. This materialist synthesis was actually perfectly in tune with the mix of Marxism, Darwinism, and Freudianism that came to signify all kinds of liberatory narratives. The new watchword could soon become: "Take control over your DNA." The old and intense fight between materialism and spiritualism took a new shape: between those who would look for genes encoding behaviors, and those who would deny such a depressive possibility. Altruism, homosexuality, and all kinds of human manners became recast as molecular operations on the stuff of life (and especially hormones and neurotransmitters).

Suddenly, you did not need to be proud of your ancestors anymore, only rejoice about "the quality of your genes." Mommy and Daddy are mere DNA providers, and Oedipus has become a fairy tale for idealists unaware of this basic truth: *there must be a molecule for everything.* The third tier of human individuation, psychic and collective, was next in the line of "problems" awaiting a molecular explanation. For the realm of the living, DNA was Simondon's "pre-individual," a mode of the universal inside each of our cells, the whole evolution transcribed into a master molecule. If individuation was Simondon's name for the universal mechanism of evolution (according to Anne Fagot-Largeault), DNA would be its material embodiment. Next, DNA could take over the noetic soul: after all, the genetic code was "the code of code," the very archetype of language.

Yes, DNA so became the closest thing ever to a common nature for the living—the common nature that is protein-based, that is.

Individuation, then, at the molecular level, has found its nearest thing to a principle in matter, as the Thomists would have thought. It is actually ironic that Thomists and molecular biologists would agree on this issue. In this scheme of things, DNA is a common nature: members of the same species share the same genes (virtual reservoir), but express (actualize) them singularly.

Well, this picture seems a bit flawed. If indeed people share 99.9 percent of their bases in the coding part of DNA, the noncoding parts are so variable in size and location that we could invent DNA "fingerprinting." By now it is common knowledge that humans are genetically identical at 99.9 percent *and* that each DNA is unique (you've seen it on *CSI*). How could it be?

Could it be that the genome (coding DNA) is a common nature, and that the noncoding part—that is, junk—as to do with individuation? And I don't mean by that the simple idea that the genome is the common nature and the junk is solely responsible for the individuation. There is, after all, variation not only in genes (at a rate of one to a hundred to one to a thousand single nucleotide polymorphisms), but also in their expression (not only among individuals, but obviously within the "same" individual through his or her development).

Could it be, rather, that DNA is the expression both of a common nature and of the singularity of a given individual? Could DNA be both the software and the junkware of life, always common and singular? Could it all be network, a web of life in every plan, molecular and molar?

Chapter 5, Lysis and Replication

Homo nexus, Disaffected Subject

> Human nature, essentially changeable, unstable as the dust, can endure no restraint; if it binds itself it soon begins to tear madly at its bonds, until it rends everything asunder, the wall, the bonds and its very self.
>
> —FRANZ KAFKA, "The Great Wall and the Tower of Babel"

"Homo nexus" is the name that I give to today's transitional form toward overman, this "new form" whose advent might make current human beings (Homo sapiens) obsolete: or, in the programmers' lingo, posthumans might eventually make Homo nexus *404 compliant.*[1]

S A Philosophical Fiction

Alfred Elton van Vogt (April 26, 1912–January 26, 2000) was a Canadian-born science-fiction author, and one of its early pioneers.[2] Born in Winnipeg, the son of a lawyer, van Vogt grew up in a rural Saskatchewan community. Without money for education (like many children of the Great Depression, his father lost a good job), he did not attend college. He worked at a series of jobs and then started writing true confessions, love stories, trade-magazine articles, and radio plays. In the late 1930s, he switched to writing science fiction, influenced by his teenage passion for fairy tales. In December 1939, he published his first SF story, titled "Discord in Scarlet," in John W. Campbell's *Astounding Science Fiction,* the ultimate science-fiction serial of all time. In the same issue appeared Isaac Asimov's first *Astounding* story, "Trends"; Robert Heinlein's first story "Lifeline" appeared a month later, and Theodore Sturgeon's "Ether Breather" a month after that. Van Vogt thus participated in the first generation of the golden age of SF in the United States.

"Discord in Scarlet" depicted a fierce, carnivorous alien stalking the crew of an exploration ship in outer space. In 1950, van Vogt incorporated the story into his novel *The Voyage of the Space Beagle.*[3] The plot of the story, in its various versions, always revolves around a "close encounter of the third kind," a malevolent one, that is. Its alien menace—Coeurl, a big, black, enigmatic, catlike creature that consumes "id" and can teleport itself through space—is matched against the human crew of the spaceship. The only thing that does not make it an unequal battle is the crew's use of a new science, called "nexialism."[4]

In *The Voyage of the Space Beagle*, van Vogt created a protagonist, Dr. Elliott Grosvenor (an implicit reference to the earliest cybernetic device, James Watt's regulator), who is the first graduate of the "Nexial Foundation." Trained in a kind of transdisciplinary science, Grosvenor is able to see the connection between many aspects of a problem that other specialists could not see because of their disciplinary training. Van Vogt defined nexialism as "the science of joining in an orderly fashion the knowledge of one field of learning with that of other fields. It provides techniques for speeding up the processes of absorbing knowledge and of using effectively what has been learned."[5]

In fact, "nexialism" is van Vogt's fictitious rendering of two of his main influences: Korzybski's general semantics and Alfred North Whitehead's process philosophy. The two were linked historically, and Korzybski acknowledged his debt to Whitehead on the first page of his masterpiece, *Science and Sanity*, when he dedicated his system, and thus his book, to the works of fifty-eight great authors, including Whitehead, "which have greatly influenced [his] inquiry."[6]

There is not much doubt that van Vogt coined the word *nexialism* on the basis of the extensive treatment of the concept of nexus in Whitehead's philosophy. In *Process and Reality*, Whitehead makes the nexus one of his central concepts, which, along with those of "actual entities" and "prehensions," describe the "ultimate facts of actual experience."[7] While actual entities, also dubbed "actual occasions," "are the final real things of which the world is made of,"[8] "prehensions" are relations among actual entities:

> Actual entities involve each other by reason of their prehensions of each other. There are thus real individual facts of the togetherness of actual entities, which are real, individual, and particular, in the same sense in which actual entities and the prehensions are real, individual, and particular. Any such particular fact of togetherness among actual entities is called a "nexus" (plural form is written "nexûs").[9]

In a realist fashion,[10] nexûs are thus "particular entities," one of Whitehead's eight categories of existence.[11] Van Vogt's fictitious rendering of Whitehead's philosophy thus proposes a characterization of a science of relations where relations (prehensions) are real entities, and not "mere" abstractions.[12] Whitehead adds that he uses "the term 'event' in the more general sense of nexus of actual occasions, inter-related in some determinate fashion in one extensive quantum. An actual occasion is the limiting type of an event with only one member."[13] This crucial aspect of Whitehead's process philosophy (or relation philosophy)[14] was picked up by Gilles Deleuze, who saw in Whitehead one of the very few precursors of his own event philosophy:

> According to Whitehead, the component element of the event is the prehension . . . a prehension constitutes an event. Or, since an event is a conjunction that corresponds to several conditions, we ought to say that it is itself a link, or, as Whitehead says, a nexus. The event, form the point of view of its composition, is a nexus of prehensions. From the point of view of its conditioning, it is a conjunction of series, from the point of view of its composition, it is a nexus of prehensions.[15]

R **The Debt and the Contract**

But "nexus" is not only a Whiteheadian concept borrowed by A. E. van Vogt. In Roman law before Justinian, a person called *nexus* or *addictus* was a quasi slave, not exactly a slave, but treated as such: he retained his personhood (slaves were not persons). The Romans had no prisons for debtors, and the creditor was the debtor's jailer. A person was called *nexus* when he was bound to a creditor and had given himself—that is, his body and his labor power—as security for his loan. In case he could not pay his debt in time, he became *addictus.*[16]

Nexus and *addictus* were thus two legal variations on the specific kind of subhumans that Romans called slaves. Under such conditions, free persons could enter the realm of *res mancipi,* of things such as "land, houses, slaves and four-footed beasts of burden" that could be owned and thus required *mancipium,* transfer of ownership, corporeal apprehension.[17] From this troublesome origin, nexus became the origin of all contract. There are various explanations of the origin of the word in Latin. For some, it seems to come from the contraction of "*neque suum*" (and not his own), as is found in this excerpt of Varro's *De Lingua Latina* (Book VII, 105):

> Manilius wrote that nexum is all that is done by bronze and scale, including mancipium; Muncius, wrote about the things which are bound by bronze and scales, excluding those given by mancipium. That the latter is truer, the word itself shows it: for this bronze that is bound by scales does not make it his own, and is thus called nexum. The freeman who was giving his labor in servitude for the money that he owed was called nexus until he honored his debt, as burdened by bronze.[18]

In the Merriam-Webster online dictionary, "nexus" has three interrelated meanings that date back in English to 1663: (1) connection, link; also: a causal link; (2) a connected group or series; and (3) center, focus.[19] Its etymology is reported to be the past participle of the Latin *nectere,* "to bind." The *American Heritage Dictionary of the English Language,* in its fourth edition (2000), gives the same three meanings,[20] but traces the Latin origin to the Indo-European root **ned-:*

> DEFINITION: To bind, tie. 1. O-grade form *nod-. a. net1, from Old English net(t), a net, from Germanic *nati-; b. nettle, from Old English netel(e), netle, nettle, from Germanic *nat-ilo, a nettle (nettles or plants of closely related genera such as hemp were used as a source of fiber); c. ouch2, from Anglo-Norman nouch, brooch, from Germanic *nat-sk-. 2. Lengthened o-grade form *ndo-. node, nodule, nodus, noil, noose; dénouement, from Latin nodus, a knot. 3. With reformation of the root, nexus; adnexa, annex, connect, from Latin nectere (past participle nexus), to tie, bind, connect. (Pokorny 1. ned-758.)[21]

Junk and nexus thus come from related semantic fields stemming from two different Indo-European roots: *yug-* and *ned-.* They provide us with our two main archetypes of the machines of the first kind: the yoke and the knitting needles.[22] Put together, these two archetypes organize the becoming of computing machines, through difference engines and Jacquard looms (second kind). With them they carry implicitly, and nowadays even covertly, their Indo-European referents to the pastoral (i.e., nomadic) economy.

Today's and hence tomorrow's Homo nexus is the connected man, the bound or tied-up individual. As the original quasi slave he is, literally speaking, (k)net-work. It is thus not surprising that the popular wisdom refers to slavery or addiction alike, when it alludes to the relationship that some of us already entertain with our communication prostheses, cell phones, laptops, and so on, but especially those prostheses that make us paradoxically (?) mobile workers (i.e., nomadic agents).

Homo nexus is thus blessed and cursed, beyond good and evil.

He is both the prophetic and the actualized figure of our future, today's face to come (as one node in a series of interfaces). For this he is blessed. His connectedness opens for him the realms that used to be ascribed to the divine only (and to a lesser degree to angels): through his connections he can reach ubiquity and omniscience, no less. This is the age-old promise bundled up with every new communication medium, and it is indeed a religious promise, the other side of a covenant: religion might come from *religare,* to bind fast; communication in this sense is a form of communion. I call this Metatron's promise: the computer, who was first a person, becomes a person again, the free person who was computed, hence a node in a distributed network of such singular entities, each its own species, like angels. The computer is born again and in this second birth he is made an angel on Earth (including its noosphere). We might have created a hole in the ozone layer of our atmosphere, but we also created trails in the noosphere, our new frontier. The computer that we know now, this box on our desk or on our laps, is but an obsolete intermediary: the future of computing is indeed organic, and DNA is tomorrow's processor. *We* are tomorrow's computer, part calculating machine, part language machine, forever dwelling in real time—that is, not chronological time or eternity but the concomitance of *aevum* and chronological time in discrete time. We are the second coming of the computer, and this is an apocalyptic coming: from this DNA you will be born again, and again. Blessed be the stem cells of Homo nexus!

But Homo nexus is also the eternal return of the pastoral slave, the credit-craving addict of the capital of the fourth kind, where blood, gold, and sand have given place to code itself qua nucleic acids, fulfilling the destiny of the living money. He is our ancestor as well as our heir, "the wound and the knife," "the limbs and the wheel" (Baudelaire, "The Heautontimoroumenos" or the self-tormentor), and as such he is cursed. Or, better said, *damned: damnatio* was the name of the solemn formula uttered by the creditor at the moment of *nexum*, when the bronze was weighted. In Rome as in 1984, *damnatio memoriae*, erasure from public memory, cancellation of any trace of a life, was reserved to the Enemies of the Empire (*down the memory hole*, says Orwell). As code itself is sold and exchanged in its own medium, as code is now the currency, the general equivalent, and the product, the object of the transaction, political economy and the political economy of the sign conjugate: *damnatio memoriae* by ubiquity. At the very moment of his conception, Homo nexus knows that he has already fallen, twice fallen, in fact. If he is human still, he owes it to this Adamic reminiscence; if he will be an angel soon, it is of the dark kind: the kind that decided to exercise his freedom and design himself (and it is no more a mere question of appearance that this design solves). Square Fall, fallen Angel and knowledge-drunk Golem. In Homo nexus all the eternal returns converge in the One singularity, who, having successfully stripped the divine of his attributes, veils himself in the emperor's clothes while dancing on his grave. The emperor is naked, design was an illusion, the emperor is dead, long lives Homo nexus! God is dead, long lives Man-God.

Homo nexus will honor all debts in one transaction. He will become Debt and the Debtor, Credit and the Creditor: a perfect simultaneity that abolishes debt and credit by being debt and credit. Homo nexus is the future of capital as biocapital: "No longer simply the attribute of a sovereign organism, life now emerges out of the connections of a network, involving an essential impropriety—it is life's habit of refusing containment that becomes interesting for biotechnology and capital."[23] An unrepentant slave who escaped servitude, *liber nexum,* is the archetype of this decoded life, in the sense of Deleuze and Guattari: "let us recall that 'decoding' does not signify the state of a flow whose code is understood *(compris)* (deciphered, translatable, assimilable), but in a more radical sense, the state of a flow that is no longer contained in *(compris dans)* its own code, that escaped its own code."[24]

For the story of the nexus is never as interesting as when he is freed at last, *untied.* Titus Livius (*Roman History,* Book II, 23, 1–8) tells such a story, which might be history or myth, according to Georges Dumézil: it is the story of an imminent war and a rebellion of the soldiers-nexi (think Buffalo Soldiers, Foreign Legion, etc.) who were freed to save Rome, and build its Empire. To Dumézil, this mythic story is "one of the rare evidences that we have on the oldest 'Männerbünde' of Italy."[25] This word, which could effectively be translated by "man-bond," gives yet another sense to the *nexum,* quite in tune with Livy's story: that of "an all-male comrade-in-armship," another scary reference to the already-too-scary *repertoire* of the overman. Scary because the thesis that Indo-European states were born out of *Männerbünde,* rather than family or market, is usually ascribed to the fascist theories of the kind best expressed by Julius Evola's *Men among Ruins.*

Deleuze and Guattari chose another way, and followed there the Marxist way—could there still be only two ways, fascist or communist, asks the European intellectual? Following Marxist anthropology (Ferenc Tökei, 1930–2000) and archaeology (Gordon Childe, 1892–1957), they consider that *the paradigm of the bond or knot is the erection of a state apparatus upon the primitive agricultural communities,* "submitting them to the power of a despotic emperor, the sole and transcendent public-property owner, the master of the surplus of the stock, the organizer of large-scale works (surplus labor), the source of public function and bureaucracy."[26] Later, they add: "this is the regime of the *nexum,* the bond: something is given without the transfer of ownership, without private appropriation, and the compensation for it does not come in the form of interest or profit for the donor but rather as a 'rent' that accrues to him, accompanying the lending of something for another's use or the granting of revenue."[27] This is the orthodox view of "Asian despotism" all right—for the heterodox kind, Wittfogel's kind, leads to quite another conclusion: sending back to back fascism and communism, understanding Stalinist despotism as Oriental despotism—but this is another story. Let us keep up with Deleuze and Guattari's flow here, because their overlooking of this nasty despotic becoming will stem our understanding of tomorrow's Homo nexus (this is the wager):

> Are there people who are constituted in the overcoding empire, but constituted as necessarily excluded and decoded? Tökei's answer is the *freed slave.* What counts is not the particular case of the freed slave. What counts is the collective figure of the Outsider . . . *The bond becomes personal* . . . even slavery changes; it no longer defines the public availability of the communal worker but rather private property as applied to individual workers . . . Private property no longer expresses the bond of personal dependence but the independence of a Subject that now constitutes the sole bond.[28]

All this leads to their redefinition of capitalism, the opposition of machinic enslavement and social subjection, and their typologies of machines/states of the three kinds that I have used previously (see chapter 4); but here I can close my loop and focus instead on the quasi slave himself, this new form of slave (living money, literally speaking) that comes with the new breed of capitalism that I have dubbed, after their work, capitalism of the fourth kind. Let us thus assume that what does count is the particular case of the freed slave, and that it has *always* counted (again, literally speaking). If "the modern States of the third age do indeed restore the most absolute of empires," if "Capitalism has reawakened the *Urstaat,* and given it new strength,"[29] the table is set for the eternal return of the *nexum* under a new guise:

> There is a unique moment, in the sense of a coupling of forces, and this moment of the State is capture, bond, knot, *nexum,* magical capture. Must we speak of a second pole, which would operate instead by pact and contract? Is this not instead that other force, with capture as the unique moment of coupling? For the two forces are the overcoding of coded flows, and the treatment of decoded flows. The contract is the juridical expression of the second aspect: it appears as the proceeding of subjectification, the outcome of which is subjection.[30]

Such is, according to Deleuze and Guattari, the scheme of the eternal return of machinic enslavement, finding its original description in Marx: "a cosmopolitan, universal energy which overflows every restriction and bond so as to establish itself as the sole bond."[31] From the imperial bond, the agrarian despotic bond of collective forced labor to all the forms of subjective, personal bonds and eventually back to the self-binding of a subject thus "renewing the most magical operation."[32] Should we, they ask, invoke "voluntary servitude" here? No, and it is the strength of the strange feedback that they mobilize here: "machinic enslavement presupposes itself," it is cause and consequence of a circular causality, the engine and the fuel of the process, beyond the false opposition of "forced" and "voluntary": it is not a question of will, only a conjunction of forces. Such is the feedback loop of the bond, the knot, the *nexum*. And yes, they are right, there is only a unique moment, with two coupled forces that conjugate. For, as a long scholarly tradition bears witness, the *nexum* is both the pact and the contract: "The oldest contract of Roman Law, the *nexum*," writes Marcel Mauss in 1923, "is already detached from the ground of collective contacts and from the system of ancient gifts that bind . . . there is certainly a bond in the things themselves, *in addition to* the magical and religious ties and to those of the words and gestures of the juridical formalism."[33] Georges Dumézil too, wrote about "this oldest fragment of Roman law, that is only known to us as stripped of any religious element . . . the oldest regime of debts, which the two words *nexum* and *mutuum* dominate."[34] Not one, but two words: *nexum*, but also *mutuum*, "formed on the Indo-European stem-root **mei-*, exchange gifts (as in potlatch)."[35] If *nexus* refers to the debtor enslaved to his creditor, *mutuum* refers to the loan:

> Historians think often as if the beginnings of Roman law were an absolute origin; but before the *aes mutuum*, before bare *aes*, there were certainly already some contracts, at least some gifts and binding exchanges, some potlatches, all that expressed the root **mei-*, and these juridico-religious acts themselves ought to be about a material thing: it is not by chance that *pecunia* derives from *pecus*; the *mutuum*, "the thing given for—mandatory—later return" (and later, "the loan"), of the Indo-European pastors who invaded the Latium must have been, most of the time, cattle.[36]

Pecunia, money; *pecus*, cattle: livestock is the original form of living money. Even in Livy's original story/myth of the Roman nexi, this was the key to the first account of the personified nexus.[37] I hope, dear reader, that you will pardon me the antiquities . . . But all the elements that I needed are now in place (all these elements, I am sure, that Deleuze considered obvious since he had been trained in this tradition). This is how, according to the tradition, "something could be given without transfer of ownership." This is how the given thing embodies the bond itself, being a part of the family handed over to another family, and let the *stigmata* be the only sign of this belonging on the body of the thing—cattle, slave—itself.

Here I beg to differ, however, with Deleuze and Guattari, and, in a lesser fashion, with Baudrillard, who all seem to consider that they somehow completed the critique of political economy. Baudrillard wrote in 1972: "today we are at exactly the same point as Marx was. For us, the critique of political economy is basically completed . . . and according to the same revolutionary movement as Marx's, we affirm that we must move to a radically different level, which permits, beyond its critique, the definitive resolution of political economy. This level is that of symbolic exchange and its theory."[38] As Feuerbach had completed the critique of religion, leaving Marx to the reinversion of religious form into critique itself, Baudrillard was then left to the reinversion of the political economy into critique itself, to overturn its ambiguous limit into the political economy of the sign itself, *the critique of the metaphysics of the signifier and of the code.* For Deleuze and Guattari, on the other hand, this sense of completion is brought upon by "the extreme perversion of the contract reinstating the purest of knots"—political economy leading, quite differently it should be said, but leading anyway, to an economy of *desire.* And when Deleuze and Guattari sought "the desiring machines" and "their arrangements and syntheses" of "the molecular machinic elements" in the "genic unconscious," and following the Marxist intuition again, "non-human sex,"[39] Baudrillard strangely concurred (for once) and found that "at the level of genes, the genome and the genotype, the signs distinctive of humanity are fading . . . we have the perpetual motion of the code, the metonymic eternity of cells."[40] Period pieces!

What if the "definitive resolution of political economy" was only rebooted by the (real) discovery of the genic unconscious? What if "the extreme perversion of the contract" was not completed by this individual interiorization? What if there was rather a molecular stratum where the contract (and the magical capture) could take place anew? And what if there was indeed an already always decoded flow at this molecular level itself: "What does it mean for a species to have the right to its own genetic definition, and thus its potential genetic transformation?" wrote Baudrillard, who instantly reminded us that "it seems that 90 per cent of the human genome is of no account. Are we going to claim this obscure part which has no apparent purpose?"[41]

You bet we will—unless it claims us!

Homo nexus shall again be the new *mutuum,* the self-loan and the sacrifice of his obsolete humanity. Homo nexus shall redefine the frontier between human and nonhuman, and perhaps even charter a third way, an escape from this anthropocentric dualism: neither human nor nonhuman (sex, desire, drive, will, freedom, rights), but alien to both, open, excluded and marginal to code itself. Already freed slave, eternally returning voluntary serf, magically captured being in his own code, Homo nexus shall necessarily resolve the human conundrum that was here from the start, and start over (reboot humanity). Slave to (narcissist) Love. In this process, design, this antiquated fifth way to know God, shall have become the only way to know man-God, the star eater, his successor. Homo nexus, in his renewed pantheism, shall put God to sleep, at last, and honor the worlds he created without God's help. And junk shall attest to these interior worlds, junk shall be the stigmata of his belonging to the family of the living, junk shall be the eternal rest (in the sense of sleep as well as in the sense of remainder) of the God principle inside his code.

Nexus juncus shall be his true name when Homo shall fade, and no family but the family of the living shall be his kind. And then again, you might ask: should we convene an ethical committee to decide if he may, and discuss the law, the contract, the patent, and the copyright notice?

O **Dis-Affect**

What about us, then, the ante posthumans, the not yet radically transformed beings, us, who live at the brink of the evolutionary leap, who might see it happen (to somebody else) in our lifetime? Are we, alas, to be overman's subhuman, his evolutionary sidekick?

To us, all this posthuman affair still sounds like a metaphor . . . and Homo nexus is nothing but its allegory: this creature of our time, for whom Super Bowl ads are targeted. *Live out loud: the latest MP3 phones (are for you).* To him, *the future is (user)friendly* . . . but to us? To us corresponds this question, asked by the most lugubrious novelist of our (closing) times: "Who among you deserves eternal life?" (Michel Houellebecq, the high priest of dis-affection, in *The Possibility of an Island*). Welcome to the new rat race! What has one to do in order to actually deserve immortality? Where should I register? asks the derelict. For Raelians, press "one"; for Extropians, press "two"; for Dianeticians, press "three"; for Singularitarians, press "four"; for a guidebook with all our options for medical tourism, press "five" . . . And be ready, the average waiting time is one generation.

And yet, we, creatures from the postindustrial age, have already been called "*living money.*" Homo nexus's literal fate (and not ours yet, for ours is still death only). Writing about us, but already feeling that his time might come soon, Pierre Klossowski wrote (and I hope that you will forgive the lengthy quote here):

> "Living Money," the industrial slave counts both as a sign guarantee of wealth and as wealth itself. As a sign she is worth all kinds of other material wealth; as wealth, however, she excludes all other demands but the one she satisfies. But actual satisfaction, her quality as a sign equally excludes. Here is in what living money essentially differs from the industrial slave (idol, star, advertising model, hostesses, etc.). The latter could not claim to be called a sign so long as she

> distinguishes between what she agrees to receive, in inert money, and what she is worth in her own eyes . . . As soon as the corporeal presence of the industrial slave enters absolutely into the composition of the assessable yield of what she can produce—her appearance being inseparable of her work—it is a specious distinction, that of a person and her activity. Corporeal presence is already a commodity, independently of and *in addition to* the commodities that this presence contributes to produce. Henceforth, the industrial slave either establishes an intimate relation between her corporal presence and the money she makes, or she substitutes for the function of money, being money herself: at the same time equivalent of wealth and wealth itself.[42]

In Homo nexus already lures the figure of the eternal return of the slave, still an allegory. And we, in the appearance of a perpetual ad that would have run amok a long time ago, always posing in front of the imaginary camera of the reality show that now is our life, its contemporary reflection, its ambiguous actualization. Ah, spectacle! Old school was "you are what you make"; passé you are what you drive (its mobile equivalent). Today you are whom you produce; you are the producer of the appearance of reality that is called "your life." You are the offered and yet invisible image of the spectacle of your intimacy. Postindustrial indeed: how far off can we drift from industrial commodity and still make money (or credit, or fame)?

Disjunctive synthesis: equally this and not-this, yet not both (Klossowski and Deleuze concur: not stupidly Hegelian), at equal distance from this and not-this. Sign *and* material good: my body, for your consuming pleasure! Alas, how childish it was still to rehabilitate the bodily trade, the capitalist pornography, the pervert intercourse . . . Ah, how delicious it was, to still linger in transgression . . . These Matter Fuckers had quite a ball, during the last century of the second millennium, challenging us for the rest of time: say I can't cheat prosperity? We were warned, it was *the* (pre)*history of an error:* "We have abolished the real world. What world is left? The apparent world perhaps? But no! *with the real world we have also abolished the apparent world!*"[43] And in a slightly updated version: *Mid-time; moment of the surest shadow; beginning of yet another longest error; next high point of humanity; INCIPIT HOMO NEXUS.*

Cases in point: My e-mail alarm rings. I just received a new incredible offer from Johnnie, in the form of a proliferating pointer to an online pharmacy and its thousands of mood-altering, appearance-changing pills. The link was accompanied by a truncated quote from the eighth chapter of an Edgar Rice Burroughs book titled *The Land That Time Forgot* (indeed). *Apathy, for the speaker rarely. Against the Python I showed them the Thermos bottle model and Mum I.*

When entered in Google, it led to a literary online encyclopedia and a *faux-blog*, with the true URL of the online pharmacy. Fed back by the loop, swallow your pill. One of the true pills they sell is called soma, as in *Brave New World*, albeit a muscle relaxant rather than a narcotic. I considered it for a while, and opted instead for a decaf Moksha.

These eloquent accessories are created to be donned by elegant females . . . I'm talking about you! You work continuously, but you don't ever spree on anything for yourself!

Life doesn't imitate art, it imitates bad television (Woody Allen). I zap medium, the I-tube always on some kind of extreme makeover. Like this New York maiden, this week's lucky candidate, who ecstatically exclaims: "Even my dog is no longer ashamed of me when we stroll down the streets of Manhattan." In a couple of generations, she shall still wear the same made-over body, and be the happy owner of a clone of her dog. *Who, amongst you, deserves Eternal Life?* I do, because I am worth it, echoes the Broadway broad. Keep on doing the strand!

The longest error, the problem of happiness (*dixit* Aldous Huxley).

. . . heard, and seen; how I had issue beheld and detest watched himself: how I listened, how much pollution per matter becomes due? monastery to bury state that Lys finally discovered tomb that the Neanderthal . . .

Should I go on, digging deep into literal junkware? Please make me out to be an idiot savant, the philologist of spam. Which spam? The kind you actually welcome? Ah, *Néant der Thallus* . . . We are to Homo nexus what Neanderthal was to us: a bad, albeit fleeting, memory, an afterthought. Our e-toys are but his transitional objects. So, how does it make *you* feel? A bit depressed, maybe? Don't despair: now you can be one with Nature:

> Prozac 'found in drinking water'
>
> Agency report suggests so many people are taking the drug nowadays it is building up in rivers and groundwater. A report in Sunday's Observer says the government's environment watchdog has discussed the impact for human health. A spokesman for the Drinking Water Inspectorate (DWI) said the Prozac found was most likely highly diluted . . . The exact amount of Prozac in the nation's drinking water is not known.[44]

The organic composition of man, as Adorno used to say, is exploding in bits and artificial compounds of all shapes and tastes.[45] Mouth plugged to the biopharmaceutical output, anus right back to the recycler, eyes wide caught in the feedback loop of the World™, fingers free to type. Inside, digestion and appropriation, information, continuous fluxion of a perpetual blood change, dark bile of *Kultur* for the lubrication of a very small engine (with a time limit and a low power ratio).

Brother Junky, say you grew up on Ritalin™ (aka methylphenidate), shook up your teenage crisis with speed (aka alpha-methyl-phenylamine) and pot (aka tetrahydrocannabinol), moved to junk (aka diacetylmorphine) and Lithium only to realize, late in your twenties, that hope exists only in dreams. Your shrink eventually told you to slow down and prescribed Prozac™ (aka fluoxetine hydrochloride). From then on . . . you started drinking straight from the faucet.

Disaffected man, the ultimate social form of Homo nexus, his best symptom, is no ordinary desperate: he is not even worth your pity. The new Pharmakon is here to take care of him. As time goes by and he sinks ever deeper into depression, therapy gets better and the molecules more efficient. So, on he goes, quietly exchanging his labor for a handful of pills plus the psychological support that allows him to reproduce his labor power—and, if he is lucky enough, his reproductive force. Sometimes, he takes a vacation and tours the world.

Dis-affected, dis-affection, in both of the French senses of the term. In English, *disaffected* simply means "discontented and resentful esp. against authority: REBELLIOUS" (according to *Merriam-Webster's Collegiate Dictionary*). The Robert/Collins French–English "super senior" dictionary proposes to translate *désaffecté* by "disused" or, in the case of a church, "deconsecrated." Here I mean by "disaffected" both "disused" and "deprived of affect." I feel that I can reasonably have such a meaning in mind when I use this word in English, because "dis-" is a prefix referring to a negation, an opposition, a deprivation, or a lack, and "affected" has (at least) the following three meanings in English: (1) inclined, disposed; (2a) given to affectation; (b) assumed artificially or falsely: pretended (*Merriam-Webster*); "affectation" is defined as "the act of taking on or displaying an attitude or mode of behavior not natural to oneself or not genuinely felt," or "a striving after" (*Merriam-Webster*). I contend that all these meanings are appropriate to characterize today's human being, this "subject without affect" or disaffected individual. Anyway, what can rebellion amount to when singularity is the new conformity?

Strangely enough, *disaffected* is the adjective qualifying the subject/patient of the process of dis-affection, when there is no such adjective for the process of dis-affectation, for this word does not exist in English (it is absent from the dictionaries that I consulted and my spell-checker seems to resent my use of it). Both processes, however, have found their way into some current sociological diagnoses:

> In this book, I try to show that control over attention leads necessarily to the uncontrollable . . . the psychic dis-identification, to which leads the control of primary and secondary identification, leads itself to a process of collective dis-individuation, that is to say a destruction of the social body itself, and generates individuals psychically and socially dis-affected, in two senses: it generates both their *dis-affection*, ruining their affective capabilities, and their *dis-affectation*—their loss of place, that is to say their ethos.[46]

Confronted with the spectacle of dis-affection, one is taken by a paradoxical desire: either kick his ass or whine with the disaffected being. *I suffer too . . . But please react!* one is tempted to say, until one realizes that the capacity for reaction is exactly what this particular human being is cruelly lacking. From melancholia of old to today's nervous breakdown and all kinds of burn-outs, the disaffected being let his morbid shadow dwell on the last lights of modernity. Not one, but millions of specters haunting the twilight of capitalism. The disaffected individual still affects us with his unbearable haecceity, though, implacable like an existential double bind: empathy or upset won't make a difference. *Room for one more inside, Sir!* I avert his empty gaze, I cross the street, walking faster, only to understand, on the other side, that I have walked on the side of dis-affection. So, all together now: Hello. My name is X and I am dis-affected.

Soon, if not yet, the disaffected will actually envy the desperate.

Today's disaffected individual is indeed beyond control, literally speaking "out of control." His affective regression is nearly terminal: after discipline-induced guilt and control-induced shame, he reverted to plain anxiety, the first and last of the affects. In his wonderful *Profanations*, Giorgio Agamben attracted my attention on a fantastic posthumous fragment that Walter Benjamin wrote in 1921:[47] "Capitalism as Religion."[48] In this fragment, Benjamin argued that "Capitalism is probably the first instance of a cult that creates guilt, not atonement" (288). Even more jubilatory is his insight that considers that the theories of Marx, Nietzsche, and Freud "belong to the hegemony of the priests of this cult" (289). For Freud, indeed, "sin is capital itself, which pays interest on the hell of the unconscious" (ibid.). For Marx and Nietzsche, he wrote:

> The superman is the man who has arrived where he is without changing his ways; he is historical man who has grown up right through the sky. This breaking open of the heavens by an intensified humanity that was and is characterized (even for Nietzsche himself) by guilt in a religious sense was anticipated by Nietzsche. Marx is a similar case: the capitalism that refuses to change course

> becomes socialism by means of the simple and compound interest that are functions of *Schuld* (consider the demonic ambiguity of this word). (Ibid.)

The translator (Rodney Livingstone) gracefully helps us in considering this "demonic ambiguity" and reminds us that "the German word *Schuld* means both 'debt' and 'guilt'" (291n2). So, indeed, capitalist debt is the actualization of the Christian guilt in the new cult of the disciplinary societies. But it does not stop here. Agamben further actualizes it for the next step, through control—and maybe even genetic—societies.

In its current state—Agamben writes in 2005—capitalism is in its "extreme phase," also called the society of spectacle. But what spectacle? Agamben answers that here is exhibited in each "thing" (commodity) itself the separation with itself: "spectacle and consumption are indeed the two faces of the same impossibility of use." Therefore, he argues, *profanation*, the operation that returns the thing from the realm of the sacred (*sacer*) to the realm of the profane, is now impossible. Capitalism makes it impossible, and "realizes the pure form of separation, to the point that there is nothing left to separate. *An absolute profanation without remainder now coincides with an equally vacuous and total consecration.*"[49] Opposing the pilgrims of old to today's tourists (the consumers of the largest industry on Earth), Agamben actualizes Benjamin's scheme with the completion of the capitalist creation of the unprofanable: whereas the former still "participated in a sacrifice that reestablished the right relationship between the divine and the human by moving the victim into the sacred sphere, the tourists celebrate on themselves a sacrificial act that consists in the anguishing experience of the destruction of all possible use."[50] In the actualization of the capitalist cult, the sacrifice has become reflexive, and guilt has turned into anguish.

Guilt indeed used to be the psychic plague, the affect of choice for the disciplinary societies; then surveillance induced shame, and control machines produced Anders's "Promethean shame,"[51] this eerie feeling that whatever we do, we will never measure up to the standard now defined by the machine. Shame is not even a question anymore: switch on the I-tube on any given talk show and you will feel as pure as can be, devoid of any more reason for shame. Guilt is now universally reproved in the whole white world. Guilt is so outdated that even the Christians do not practice it anymore—the Catholics even changed their vocabulary to turn "confession" into "reconciliation." So, anguish is the order of the day. The *angst, l'angoisse,* rather, that goes with the constant production of the great Improfanable, an absolute separation itself, eternally reproduced in the plastic echo of our digital worlds, reincorporated into our own *flesh*, coiled into each of our cells. And yet, after there is no remainder left, there is still remaining junk. And what remains exactly to be sacrificed, then, from a time when God was not exactly dead ("hold back on your tears," as Foucault would say), but rather "incorporated into human existence"[52] to a time when Man-God is soon to become this human experience? Agamben concludes his praise by assigning the ultimate profanation to the "next generation": "the profanation of the unprofanable."[53] Through the twin figures of language and pornography—today's best symptoms of the absolute unprofanable—he even shows it the way of its ordeal.

Man-God's self sacrifice: offered, body and soul, on the altar of the machine of the fourth kind.

P **Promethean Angst**

In the original Greek myth, guilt, shame and anxiety are linked through the figure of Prometheus, the untying Titan.[54] Prometheus's shame is ultimately caused by his brother Epimetheus's guilt (his *double* fault) and causes Pandora's anxiety (the reverse of hope). Because of his brother's forgetfulness and lack of consideration for human beings, Prometheus "stole from Hephaestus and Athena practical wisdom (*entechne sophia*) together with fire—for without fire no man may acquire or make use of this—and he bestowed them upon man."[55] Aeschylus insists that "all manner of arts men from Prometheus learned": building, counting ("number, the most excellent of the inventions"), writing, medicine, divination, and so on. Most important for my thesis, Prometheus is also the tying god, *the Father of junk:*

> I was the first that yoked unmanaged beasts,
> To serve as slaves with collar and with pack,
> And take upon themselves, to man's relief,
> The heaviest labour of his hands: and
> Tamed to the rein and drove in wheeled cars
> The horse, of sumptuous pride the ornament.[56]

What did Prometheus get for such a unique compassion for humans among gods? An undying shame: "These manifold inventions for mankind I perfected, who, out upon't, have none—No, not one shift—to rid me of this shame." And the chorus confirms: "Thy sufferings have been shameful."[57]

His shame, after a correction by Zeus, became *our* shame. For Prometheus had given men the practical knowledge, but they still lacked "citycraft, of which warcraft *(polemike techne)* is a part": they lacked of justice in the governance of mankind, and "committed injury *(adikein)* one upon another."[58] Whereupon

Zeus, coming back to better feelings for the unfortunate men, sends them Hermes "with justice *(dike)* and *a sense of shame (aidos)* to bring order to their cities and common bonds of amity."[59] Zeus ties men with shame, Prometheus ties up their cattle (and thus their slaves) and unties them, and for this he is, in turn, tied up to a rock. This, he claims, he deserves and likes better than the servitude that ties Hermes to his father Zeus.[60] Ties, knots, bonds, Nexus.

And what did human beings get for such a stack of stolen gifts? They became mortal. Pandora is created, according to Hesiod, to punish them with mortality (and outdo Prometheus): Hephaestus crafts her form and all the gods give her an attribute or talent (her name means The All-gifted). She is a trap, a *dolos,* a deceptive gift that echoes a sacrifice that went wrong. She has the appearance of a goddess but the morals of a bitch (*dixit* Hesiod); she lies and deceives, she consumes food and men, she is everything wrong in a woman (of course). She carries a gift from Zeus, a jar *(Phitos)* containing all the *daemons,* the bad spirits, the ills: Pandora's box. [61] Today's all-reigning anxiety is the only affect we still feel just before opening Pandora's box, and this box, as in the myth, is no box at all, it is a jar, an amphora, an artificial uterus.[62]

In anxiety, said Lacan, "the subject is seized, concerned, interested in his most intimate."[63] The anxiety of today's disaffected individual is indeed *in between* his desire for Homo nexus as tomorrow's Other and his enjoyment *[jouissance].* In his most intimate fiber, the disaffected subject knows that in his *jouissance* he will never know Homo nexus (in the biblical sense or not), but only his remainder *[reste],* a partial, transferable, or transitional object: our e-toys, databases of the future. Desire for Homo nexus has passed the disaffected anxiety, passed the most troubling question: can we, as a species, outdo both ourselves *and* our machines? Can we be the designer of a being superior both to us and to our technology (our *Technē Sophia*)? Can we be God and the Fallen at once? Or, in Gnostic terms: Can we be the Demiurge, for a change? For Homo nexus is no simple Other, not even the Generalized Other who is still our *fellow* creature: he is just *our* creature, as much as he is our new machine. The remainder here takes all the remaining symbolic space.

The anxiety of birth, square inversed: *Incipit Homo nexus.* Homo nexus makes us relive the trauma of our birth (the first source of anxiety), a reversed and intensified trauma. His imagined birth is the reversed image of Pandora's birth: while the gods created her to punish us, and "balance the (deceptive) gifts," we long to create him to punish the gods (the metaphysics equivalent of *lex talionis*), and cheat them out of immortality. Back to square one, when there were only men (*anthropoid* = *andres*). His birth, out of the womb of today's capitalism, will make us, the disaffected individuals, the transferable, fallen objects to which he is/will have been attached: the machine of the fourth type is Homo nexus's matrix, his artificial uterus; he is linked to her by a networked umbilical cord. No wonder, then, that we feel like the last organic remainder of this whole business, its *placenta:* "it is the existence of the placenta which gives to the position of the child inside the body of the mother its character of a parasitical nesting."[64]

Homo nexus is the result of our collective desire for an extrauterine pregnancy.[65]

Homo nexus is our last hope, still trapped in Pandora's box. In the Greek myth, once the lid is opened and all the ills are freed, only one remains in the jar: Elpis, often translated by "Hope." But how on earth could hope be a daemon, a plague? Some have proposed instead to translate Elpis by "wait" (as in Spanish, where there is no difference but only one word for the verbs "to hope" and "to wait," *esperar*); some others have said that it should be translated not by "wait" in general but by the "expectation of the worse" (which is why Elpis accompanies the daemons). *Start worrying. Details follow.* Jean-Pierre Vernant, to whom I leave the last word on this subject, rather considers her as "the (uncertain) horizon of the future":

> If, as in the golden age, human life consisted only of goods, there would be no way to hope for anything else than what one already had. If life was without remedy, all given to evil and unhappiness, there would not be any room for Elpis. But since evils are henceforth inextricably mixed with goods without any chance to foresee exactly what tomorrow will bring, one is always waiting, fearing, and hoping. If men disposed of Zeus's infallible prescience, they would have no need for Elpis. If they lived confined in the present, without any knowledge of, or interest for, the future, they would still ignore Elpis. But stuck between Prometheus's lucid foresight and Epimetheus's unconsidered blindness, oscillating between both without being ever able to disjoin them, they know beforehand that pains, diseases, and death are their inevitable lot and, ignoring which shape unhappiness will take for them, they only recognize it too late, when it has already struck them.[66]

So, to recapitulate: *Espera Schuld.* It's a boy! It's a girl! It's an androgynous alien! It's an asexual creature, angel or daemon, who knows? It's Homo nexus.

In the meantime, the prospective parents are very anxious. Some say it is because of their bad habits. Everybody is a user these days, and everybody seems to overconsummate. But dis-affection is no alienation: the disaffected individual does not rebel against his social or economic condition. Nowadays, he is merely disenchanted, and rebellion is but another product on the cultural shelves. The welfare society has cruelly offered her the mirror of its own narcissism and trauma, thus putting at risk her individuation. Somehow, the collective dis-individuation crisis diagnosed by sociologists and philosophers, from Stiegler to Bauman, looks like the symptom of a generalized child neurosis. Peter Sloterdijk assigns the origins of this phenomenon to Nietzsche: not to Nietzsche's thought or ideas, but to his own person. He considers him the "trend designer" of today's breed of (paradoxical) individualism. According to him, "Nietzsche understood that the inescapable phenomenon that was to happen would be the necessity to distinguish oneself from the masses. By a direct intuition, he understood that the fabric of the future would be in the individual demand: to be different and better than the others, and in this demand precisely, be like everybody else."[67]

Singularity is the new conformity. Hell, it is the new market, and one of infinite potential at this—OK, it might be slightly exaggerated; say it is a market niche of six billion and rising. Business gurus understood it at the end of the millennium, and "personal branding" was born. To the nagging question "What makes you different?" Tom Peters answered: "It's this simple: You are a brand. You are in charge of your brand. There is no single path to success. And there is no one right way to create the brand called You."[68] Peters did not take the precaution of opening a dictionary. If he had, he would have realized that if today's meaning of a "brand" is "a trademark or distinctive name identifying a product or a manufacturer," it used to refer first to "a mark indicating identity or ownership, burned on the hide of an animal or into the flesh of criminals with a hot iron," and, more generally, "a mark of disgrace or notoriety; a stigma." He would have realized that the meaning of a "brand" used to be the mark that characterizes the slave, when to him, it means just the opposite: "Instead of making yourself a slave to the concept of a career ladder, reinvent yourself on a semi-regular basis." I guess unemployment is high enough for this spectacular inversion to work: now you are the brand, and you are the iron! In this global market where every consumer is potentially his or her own brand and where marketing has become viral, shopping has arguably become "the last remaining form of public activity" (Rem Koolhaas, see chapter 6). If you do not brand yourself, soon somebody else will brand you: you will even pay extra money for that. Brand yourself or be branded!

Here again, the basic reflex behind such a phenomenon is job anxiety, fear of losing your job and not finding work again. Losing your job for a while would be OK, I guess; some might even call it "a vacation." What's terrible nowadays indeed is the possibility that if you lose this job, you might not find any other job at all. And the worst is that the more people in your situation (jobless), the less chance you get of finding another job. So, what you are really anxious about is more the fact that you might have to stick forever to this particular job . . . Here again, "Anguish is not," as Lacan writes in contradiction to Freud, "the signal of a lack, but of something that you must manage to conceive of at this redoubled level of being the absence of this support that lack provides":[69] lack square, lack of a lack.

This is the master equation of today's disaffection. "One is disaffected" means that one is afraid to lack the support that absence provides: renew desire, make presence more enjoyable. This is particularly clear of all forms of addiction that signify more clearly than any other psychosocial phenomenon the generality of disaffection. The addict lives in the perpetual anxiety not of lacking of the product of his affection, but really, not to ever lack it, and therefore to get used to it, to need to increase the dosage, to get deeper and deeper into "the algebra of need" (William Burroughs).

Now, if we get back to the announced posthuman, to the promise of the inception of Homo nexus, we realize that we do not have to expect the worst to be anxious about it. Quite the contrary, here again we fear success above all. For we feel that his coming will be our obsolescence, and there is one level in particular where this obsolescence is quite meaningful: the end of sexual generation as we have known it so far.

Most of today's commentaries on the soon-to-come posthumanity have stressed this point, and there is not a month without an announced spectacular scientific progress in this sense. In July 2006, for instance, a German–British team announced that they had managed to create synthetic sperm out of embryonic stem cells and used them to produce live offspring for the first time.[70] The first newspaper article that delivered the news to me, from the Montreal-based *La Presse,* concluded with the following remark: since embryonic stem cells are by essence undifferentiated, "one can think about creating sperm from stem cells of a female embryo."[71] It happens that in the study discussed, the embryo used were male, but this was totally irrelevant to the outcome. This extraordinary "solution to male infertility," as most papers covering the news were prone to say, could also mean that soon enough, males would not be needed anymore to "produce viable offspring." By an extraordinary coincidence, the London *Times* reported this same week that in a proposal of the health minister amounting to "the most radical shake-up of Britain's embryology laws for 16 years," "a child's need for a father will no longer have to be considered by clinics before they provide IVF or sperm donation services to single women and lesbians."[72] In less than a century, men had moved from being the sexual organ of machines to not being needed at all—quite an interesting (fictitious) trajectory.

As we have seen, Samuel Butler already had this intuition at the end of the nineteenth century: during the industrial revolution, human beings slowly transformed into the missing sexual apparatus of the machine:

> Surely if a machine is able to reproduce another machine systematically, we may say that it has a reproductive system. What is a reproductive system, if it be not a system for reproduction? And how few of the machines are there which have not been produced systematically by other machines? But it is man that makes them do so. Yes; but is it not insects that make many of the plants reproductive, and would not whole families of plants die out if their fertilization was not effected by a class of agents utterly foreign to themselves? Does anyone say that the red clover has no reproductive system because the humble bee (and the humble bee only) must aid and abet it before it can reproduce? No one. The humble bee is a part of the reproductive system of the clover. Each one of ourselves has sprung from minute animalcules whose entity was entirely distinct from our own, and which acted after their kind with no thought or heed of what we might think about it. These little creatures are part of our own reproductive system; then why not we part of that of the machines?[73]

Deleuze and Guattari once commented on this very excerpt, only to conclude that since then already, the distinction between the mechanical and the vital made no sense at all, and that "it becomes immaterial whether one says that machines are organs, or organs, machines."[74] Moreover, they insisted that the essential here is that "once the structural unity of the machine has been undone, once the personal and specific unity of the living has been laid to rest, a direct link is perceived between the machine and desire, the machine passes to the heart of desire, the machine is desiring and the desire, machine."[75] And, they added: this is no metaphor. In this perspective, then, Homo nexus, this posthuman entity to be, *is* a living machine. The laying to rest of the personal and specific unity of the living, applied to the case of the human as sexual apparatus of the machine, now means: a gamete, sperm, or egg, a transferable part of a human being, is as much a part of the machine of the fourth type as any other receptacle of code. It does not need anymore the fiction of a whole person to be part of a production process. In return, of course, one might then feel that the human person, at least symbolically, has been severed of this "organ": or, in other words, today's disaffection alludes to the castration anxiety that we feel with respect to Homo nexus.

This point is hardly new. In a blazing aphorism, James G. Ballard once wrote that prosthetics is "the castration complex raised to an art form."[76] In her review of one of the—if not the—most influential accounts of the upcoming posthuman in literature, Katherine Hayles's *How We Became Posthuman,* Linda Brigham characterized Hayles's move from the presence/absence pyschoanalytic dialectic of Freud and Lacan to a cybernetic dialectic of pattern/randomness (a paradigm shift). She wrote:

> The governing anxiety in the new configuration becomes whether one is human or not, rather than whether one is male or female, and the posthuman becomes an updated analogue of castration anxiety. The paradigm shift intertwines with an economic shift: no longer is wealth a function of possession of discrete valued things (including one's body), but a function of access, facilitated by codes. Indeed, information, in its literal sense "informing" those who access it, *is* capital, from which subjects, in their ungrounded mobility, become less and less distinct. Hayles cites William Burroughs's description of dope-dealing in *Naked Lunch* as the practice of new capitalism: the "junk merchant does not sell his product to the consumer, he sells the consumer to his product. He does not improve and simplify his merchandise. He degrades and simplifies the client."[77]

Katherine Hayles, however, did not go this far and only claimed, "mutation is the catastrophe in the pattern/randomness dialectic analogous to castration in the presence/absence dialectic. It marks a rupture of pattern so extreme that the expectation of continuous replication can no longer be sustained."[78] In other words, she uses paradigm in the Greek sense of "example," but the structure remains unchanged: the duality presence/absence is replaced by pattern/randomness. Actually, one can even claim that the second duality is but a special case of the first, pattern/randomness being merely the presence/absence of order: it is thus a move down, rather than up, the abstraction ladder. In Hayles's chiasm, "mutation" replaces "castration" as the catastrophe preventing replication, "the visible mark that constantly testifies to the continuing interplay of the dialectic between pattern and randomness, replication and variation, expectation and surprise." Nowhere does Hayles write about "anxiety" and nowhere does she equate the posthuman with castration anxiety. But Brigham's exaggeration, as well as the connection with Burroughs's junk, serves me well here. For Hayles does link junk to mutation. Following the quote reported by Brigham, she concludes: "the junkie's body is a harbinger of the postmodern mutant, for it demonstrates how presence yields to assembly and disassembly patterns created by the flow of junk-as-information through points of amplification and resistance."[79]

Mutation, stemming from "the example of the genetic code,"[80] is the nexus of Hayles's demonstration, her master concept. It is, of course, in quite a different direction that I want to go here. Junk is in no way reducible to information, and mutation is quite a minor catastrophe: the presence/absence dialectic is a model one should altogether give up here. Junk is in itself, beyond presence and absence, pattern and randomness, the figure of the potential, the always recyclable. Isn't recycling today's name for the eternal return? If Homo nexus alludes to a castration anxiety of today's disaffected individual, it is because "*castration is the basis for the anthropomorphic and molar representation of sexuality,*" it is because "castration is the universal belief that brings together and disperses both men and women under the yoke of one and the same consciousness, and makes them adore the yoke."[81] Can we turn this around and make it a positive aspect of human life? Can junk be this yoke, an open one at that, one that would link us to any form of life, in the respect of both life and human difference?

For, in the end, species and hope share the same root (surprise! it is a rhizome, after all!): *-spek,* which also gives expect, and gave *Spes* the Latin form of Greek Elpis, the last evil gift remaining in the jar. What type of Hope could generation disconnected from parenthood mean? In which way could it also mean parenthood disconnected from generation?

Suddenly, it dawned on me that my own life pointed toward a different perspective on the future of generation, one about the essential lack of parenthood of future generations. I understand now that the future of generation might be built on a profound desire for a lack of parenthood. But maybe parenthood was always already a lack? Do babies without genitors relate to bodies without organs? Is there some positivity in envisioning babies with parents but without genitors? Science fiction described such a possibility in the past, but the result was not a pretty picture:

> "In brief", the Director summed up, "the parents were the father and the mother." The smut that was really science fell with a crash into the boys' eye-avoiding silence. "Mother", he repeated loudly, rubbing in the science; and leaning back in his chair, "these", he said gravely, "are unpleasant facts. I know it. But then, most historical facts *are* unpleasant."[82]

One thing seems obvious: the relationship between parenthood and generation will change. Babies will be (in part or in parts) generated, that is, produced, by machines of the fourth type. But these machines are complex assemblages, of which humans will still be a part (machines, not mechanisms). What kind of a part seems to be the question: slave/subject/user/product? Semen provider (genetic stock), uterus for rent, template (model to clone and/or alter): transferable components or parts (i.e., enslaved). Test-tube stem cells, retroviral embryos, enriched genetic stock, organs without bodies: full products. Parents, genitors, engineers: socially subjected/users to/of machines of the fourth kind. Socially subjected users, as in social addicts.

Today's science fiction broadened and updated the Huxleyan perspective, and the picture did not, at first sight, look more promising. In Michel Houellebecq's—another spam interruption as I type this, Oh so timely: *The office of God and the. IntentsIf I had given you this at. Ever in so true a flame of. T!How might one do sir to* flickers on my screen. And not even an ad, a URL, a sign of interest . . . this one coming from a certain Veronica Shepherd commercially located at shesaidlovely.com. Ô Veronica, bringer of victory, wiper of his face with the veil of fiction, be my shepherd!

In Michel Houellebecq's published works, I was saying before the interruption, the incoming posthumanity is looming, "a new species, asexual and immortal, beyond individuality, separation, and becoming."[83] Houellebecq even comments on *Brave New World* in his own novel, where his protagonist reflects on Huxley's "optimism," only to conclude:

> Sexual competition, metaphor of the mastery over time through procreation, has no more raison d'être in a society where the dissociation sex/procreation is totally accomplished; but Huxley forgets individualism. He could not understand that sex, once disassociated from procreation, remains less a pleasure principle than a narcissist differentiation principle.[84]

Houellebecq's solution is radical (it is a final solution, not a cure, as Père Ubu would say). It is not only the end of sexual generation, but the end of sex. Under another form (cloning) it is also the central issue of his latest novel so far, *La possibilité d'une île.* In both novels, it is a technoscientific solution, coming from biological research: "*the mutation will not be mental, but genetic.*"[85] In his interviews, Houellebecq confesses a belief in science only, even if he sometimes seems to deplore the end of religion. Accordingly, Houellebecq's literature does not need a hero, or even an antihero: it is concerned with the ordinary individual, in a kind of small-time sociobiological narrative. It is thus ordinarily that confronted with the unavoidable disappearance of the concretization of sexual intercourse in anything else than narcissist gratification, his ordinary immortal posthuman individual is terminally neurotic. And it is also only ordinarily that this neurosis, over time and instances of the "same individual," will lead to defection, that is, the return to "savagery" . . . and mortality.

In Houellebecq's fiction, even posthumans eventually aspire to death, as a final cure for their ultimate disaffection. Suicide, wrote Camus in *The Myth of Sisyphus,* is "the only truly serious philosophical problem":[86] it might remain as such for a while. Maybe Man-Gods too will die of laughter one day.

Strangely enough, there is still hope in this picture.

("One must imagine Sisyphus happy.")[87]

Chapter 6, Tail Again

Presence of Junk

> The most prudent and effective method of dealing with the world around us is to assume that it is a complete fiction.
>
> —J. G. BALLARD, *Crash*

Here I tie some of the last knots: junk is a name, which was given to this unknown part of DNA; but junk is more than a part in the compound expression "junk DNA." Junk is the alternate name of our world, the binding principle that holds it together. Junk is the cement of our cultural experience, the fractal principle that unifies our most intimate fiber (DNA) to the cosmos (space junk) and everything in between: what we ingest (junk food), where we live (junk space), what we trade frantically (junk bonds), our communications (junk mail), our (more or less) recreational drugs (just junk). Junk can be adequately used to describe any significant cultural experience in today's global culture: as everybody knows, TV is junk (hence reality TV), music is junk (recycled, sampled, etc.), movies are junk (especially romantic comedies: secondhand affects, i.e., sentimentality sold by the pound), art is junk (especially contemporary art, to the point that it should be renamed *junk art*—after all, *contemporary* seems to have outlived its shelf life, since Duchamp at least).

So: junk is the order of the day, and we'd better find some redemption in junk. Some paradoxical form of peace, that is.

My question for this last chapter is, *How did this happen?* How, and when, exactly, did our culture turn to junk? Or, in other words: when did we actually last create some radically *new* cultural experiences? And when did we instead start to recycle culture with the appearance (the *glitter*) of the new?

First, let me insist one more time on this: junk is not trash. I am not talking about trash culture here. I do not equate junk with trash, waste, garbage, or refuse. I mean junk, and junk only. Trash, garbage, waste, or refuse might very well be the fate of most junk, but it is not its necessary destiny, and the landfill is not the only way for junk to end up. Junk is one step ahead of waste, although this step might always be pure potential. There is an affect in junk, even if this affect might be the last remaining before disaffection: junk might look disaffected (like a derelict industrial space), but we still feel attached to it. And it is because of this attachment, of that affect, that it might not be totally ludicrous to look for redemption, peace at last, in junk.

If today's subject is indeed this disaffected individual, the addicted consumer, the affect-less social android, the political schizoid, today's junk culture is her playground, the figure of his collective landscape. If today's subject is to escape the terminal identity, the nihilistic black hole of total dis-affection, it thus might be thanks to something found in junk. Yes, indeed, there might be peace in junk.

H **Stigmata, or the World Dick Made**

This, I think, is the most crucial message from the mastermind who actually created it all: Philip K. Dick (PKD hereafter). For the only answer to the question "How did this happen?" could very well be: *when PKD said so.* This lousy, paranoid, speed-intoxicated pulp writer *actually created* this junk world, or, in his own words, *remembered* it first. PKD first saw through the iron cage of reality, got the first glimpse of the final *anamnesis.* PKD gave its contemporary name to Nietzsche's intuition of the eternal return: junk (or, in his own words, *kipple*). PKD, actually and practically, that is, concretely, transvalued all values, moved us all beyond good and evil, into junk. Yes, PKD is the only prophet of junk, whose word became world. From the power invested in him by the Logos, he actually created this world. He felt it in his bones and in his mind, and he recognized it like some long-gone impression, like somebody who would wake up from a long cultural coma (and this coma was named *modernity*).

PKD woke up to this world, named it, and thus made it: it thus became the world PKD made. Our world. His impossible biographer understood perfectly: in this world, *he is alive and we are dead.*[1] We are the zombies dwelling in the world PKD made: *junkware in VALIS.* The absolute irony of this is also its most amazing grace: by this very fact we are potentially saved. Because: only the lost can be found again, only the last can be first again, only the dead can rise again. Only the sinner can be washed from his or her sins. This is what redemption actually means, is it not? By this I do not only mean individual salvation, although it means that too:

> We appear to be memory coils (DNA carriers capable of experience) in a computerlike thinking system that, although we have correctly recorded and stored thousands of years of experiential information (knowledge, gnosis), and each

> of us possesses a somewhat different deposit from all the other life forms, *there is a malfunction—a failure—of memory retrieval.* There lies the trouble in our particular subcircuit. "Salvation" through gnosis—more properly anamnesis (the loss of amnesia)—although it has individual significance for each of us—a quantum leap in perception, identity, cognition, understanding, world- and self-experience, including immortality—it has further and more truly ultimate importance for the system (structure) *as a whole,* inasmuch as these memories (data) are needed and valuable to *it,* and to *its* overall functioning.[2]

No less, indeed! These few lines are the key to *Junkware.* In this intuition, PKD provides the exoteric core of my own writing. There is no pessimism here, since a malfunction begs for a fix. Aren't we, after all, the fixers of things (even if, as we shall see later, things are too often, our fix)? PKD got first this contemporary intuition: *redemption is an information retrieval problem* (just as life *is* a software problem; see chapter 2).

Let us postulate three subunits to this "computerlike thinking system," as in a von Neumann architecture: computation, control, and memory. The initiatory hack was to put the application in the memory subunit when not in use, and assign a specific kind of memory to it when in use. Dead memory is read only, where programs are stored (i.e., as data); random access memory is where the application migrates when computation is needed. In these terms (and it is no metaphor), today's problem is: to what kind of an application is junk for? If "garbage in, garbage out (GIGO)" is given, then what if junk in? The answer: garbage out, or else. What else? PKD answered: else is everything that was here from the start, before the bootstrap. Else is what else does: junk is the mother of anamnesis.

But, most important, PKD answered: the malfunction that causes the information retrieval problem is built into the system: it is a *designed* malfunction. Yes, PKD answered: the designer is a malevolent entity, a bad demiurge. PKD's answer was a Gnostic answer, it became today's Gnostic gospel. Since then, culture has turned Gnostic, PKD's way: junk culture.

Since then, that is: since the *Three Stigmata of Palmer Eldritch.* Since 1964, or maybe 1963, if one wants to take into account the time of conception. In *Stigmata,* in the overpopulated late-twenty-first-century world (including its Martian space colonies), dis-affection can be endured through the use of a drug, Can-D, which enables its user to immerse in a shared virtual environment mediated by miniature dolls and layouts. When industrialist Palmer Eldritch returns from an interstellar trip, he brings with him a new drug, Chew-Z, which is far more potent than Can-D, but threatens to plunge the world into a permanent state of drugged illusion under his control. The story, in PKD's words, "consists of a war between Palmer Eldritch (who is absolute evil) and Leo Bulero (who is not exactly 'absolute good' but rather the benign form of nonevil life with which we are daily acquainted). In a sense, the novel depicts relative good attempting to combat absolute evil—and in the end the relative good, in the form of Leo Bulero–triumphs."[3] Leo Bulero is the seller of Can-D and the "Perky Pat" layouts, thus a "drug dealer": so much for "benign nonevil"! Palmer Eldritch was at first a man, another drug dealer, but of cosmic proportion. Whereas Can-D proposes a consensual alternate reality, his product, Chew-Z, offers eternal life. This is allegedly procured through the mediation of a translation agent extracted from a lichen found on Titanian. In her analysis, Katherine Hayles notes that "if Can-D points to the sweetness of this illusion [information's promise of a realm of effortless pleinitude], Chew-Z points to the scarier possibility that, instead of a person consuming the drug, the drug will consume the person."[4] PKD gave the correct key to interpret this: switch from the metaphor of the drug to that of the viral infection:

> STIGMATA is a Satanic Bible: the novel describes the Pattern proliferating itself in, on & through humans. By a study of STIGMATA one can understand transubstantiation, which was my source & theme (my intent). It's even stated in the novel that Eldritch is the xtian God . . . But this is not an occluding, toxifying "virus"—it is an antitoxic, de-occlusive.[5]

Philip K. Dick uses the word *stigma* in its archaic meaning: "A mark burned into the skin of a criminal or slave; a brand." This meaning is the closest to the etymology of the word: "Middle English stigme, brand, from Latin stigma, stigmat-, tattoo indicating slave or criminal status, from Greek, tattoo mark, from stizein, stig-, to prick."[6] Convicts or slaves, those who bear the stigmata, lose their identity: they become numbers, objects. This mask is not their own mask; they have lost their singular persona. From the Roman origins, this is what characterizes a slave in the first place: *servus non habet personam.*

In the novel, the stigmata are at first prosthetic replacements/enhancements that Palmer Eldritch procured to his body after some rather unfortunate accidents, *prior to his trip* and his encounter with the alien entity: artificial hand or arm, jaw, and eyes.[7] They exist because Palmer Eldritch's body was first diminished. But, after the infection, they become the sign of the infection, its recognizable form. Whoever is infected wears them, and not, as Katherine Hayles believes, as a marker that they are in Palmer Eldritch's world, where he "makes the rules and infects all subjectivities with his alien identity."[8] No, there is still only one world, and one is really infected. The alien entity, the Proxer, did not create the stigmata—it only borrowed them, as its *only way to perpetuate itself in this reality.*[9] Chew-Z may be the mediator of its existence, but one still has to accept the possibility of its existence for it to exist in this reality. One still has to consume it, at least once, for it to consume one. One has to choose alienation, blurred reality, and despair. That is what Leo decides to believe, and I guess that it is also why PKD held that this "very ordinary, somewhat vulgar human being" eventually defeats Palmer Eldritch, the evil infection. Leo's act of faith in humanity is an act of faith in freedom of choice.

The alternate reality that Palmer Eldritch offers with Chew-Z requires a suspension of disbelief. This drive toward illusion might be justified by the highest promise—eternal life, no less—but it is still, as Leo realizes, a promise. To take on this promise, one must first feel discontent with this life, with this world. In this world, free will still exists, even for precogs. Theodicy's classic solution is: evil exists because God, in his goodness, made a free creature. Preestablished harmony requires it. This world is the best of all of possible worlds, because its creatures are still free (to make it the worst).

However, PKD modified the classic Leibnizian answer to the problem of theodicy. In *Stigmata,* one cannot totally escape the infection, for evil is constitutive of this world. Free will, in this version, is relative, constrained. The person is at stake. For Leibniz, all sins originate in errors (bad choices). For PKD, there are only bad choices in this world (again, even for precogs), because it was created by an evil god (the demiurge). In this resides his Gnosticism, in his confusion between God's *will* and God's *intellect* (understanding), that is, in the very notion of eternal ideas (what anamnesis is supposed to bring back, since Plato). In other words, anamnesis will bring back eternal ideas, closer to God's intellect, but not to his will. As long that you ascribe a drive (a will) to power to God himself, you are bound to Gnosticism, and you need a demiurge. There are, however, still two alternatives possible once you have made this choice, and Erik Davis is right to notice that PKD is ambivalent at this level too:

> Like the Gnostics of old, Dick flip-flopped between viewing the demiurge and his archons as evil, or as aberrant and selfish products of their own ignorance and power. The difference is crucial: the Manichaean notion that good and evil are absolute principles sucks you into a harsh and rather paranoid dualism, while the other, more "Valentinian" mode of gnosis opens into a continual transformation, an awakening that's always on the fly. For the Valentinians of Alexandria, the moment of transcendence is not an E-ticket out of here but a signal fed back into the maze of the churning world.[10]

This difference may be crucial, but both alternatives rest on the same principle. Whether PKD hesitated between paranoid dualism or the possibility of anamnesis was well documented by his biographers, not to speak of his own work. In the second case, however, by far the most profitable for his psyche and the source of all his hope, the crucial point that Davis makes is that *transcendence is fed back to immanence.* If, as Erik Davis had it, PKD's Gnosticism fed transcendence back in the immanent "churning world," it means that glimpses of God can be found in the creation of the evil demiurge. No one, I think, has understood and expressed this idea better than Alexander Star, in a cover essay for the *New Republic:*

> Dick's fallen worlds are not, to put it mildly, happy places. And yet they are at least partially redeemed by fleeting glimpses of a hidden god. "Trash" and divinity, Dick believed, were intimately linked. In an Exegesis entry, he wrote: "Premise: things are inside out . . . Therefore the right place to look for the almighty is, e.g., in the trash in the alley.". . . Carrying on a distinctly American visionary tradition, Dick proposed that God preferred industrial waste to holy sanctuaries. In its spiritualization of the coarse and the vulgar, Dick's demotic Gnosticism unexpectedly echoes Emerson, or Whitman, or even Melville. He sought a kind of urban sublime, looking for shards of divinity in piles of junk.[11]

Junk, then, appears as the only potential site to recover transcendence in the iron prison that this world came to mean to Philip K. Dick. This characterization opens a new realm of hope, albeit, as always in his fiction, in an ambivalent manner.

M **Kipple**

In the first released version of *Blade Runner,* the opening crawl situates the plot around the confrontation between "replicants," quasi-perfect androids, and "blade runners," the bounty hunters in charge of "retiring" them:

> Early in the 21st Century, THE TYRELL CORPORATION advanced Robot evolution into the NEXUS phase—a being virtually identical to a human—known as a Replicant. The NEXUS-6 Replicants were superior in strength and agility, and at least equal in intelligence, to the genetic engineers who created them. Replicants were used off-world as slave labor, in the hazardous exploration and colonization of other planets. After a bloody mutiny by a NEXUS-6 combat team in an off-world colony, Replicants were declared illegal on earth—under penalty of death. Special police squads—BLADE RUNNER UNITS—had orders to shoot to kill, upon detection, any trespassing Replicants. This was not called execution. It was called retirement.

After a long aerial shot zooming in on the Tyrell headquarters, an immense Mayan pyramid-like building, the action starts with a scene between two characters, a blade runner and a replicant, "inside the Tyrell Corporation locker room." The scene represents an inverted Turing test, the Voight-Kampff test ("the Eye" in the original screenplay). The result of the test is a matter of life or death: supposedly testing the empathic reaction of the subject through a measure of the dilatation of his pupil, the test ends with its retirement in case of failure (i.e., in case a lack of empathy is detected). Well, not exactly *life and/or death,* maybe, because, as a (human) character realizes in the novel, it cannot be death since androids (replicants) "are not *actually* alive."[12] Or are they? Whatever the answer to this metaphysical question may be, here they are, facing each other. Where are they? In "a large and humid room, whose walls are neatly stacked with rows of salvaged junk." The technological center of the Empire, the very core of the colonizing process, the locker room of the design factory of the replicants, is a room full of junk.

Junk is right there at the heart of the Empire, for act I, scene I. The whole movie actually portrays a world of junk, a junked world, a world where the Environmental Protection Agency and the homicide squad have merged.[13] A world where everything left—buildings, people, pets, and even affects and ideas—is junk.

This is perhaps the clearest in a particular character of the novel, John R. Isidore. In *Do Androids Dream of Electric Sheep?,* Isidore is introduced as a driver for a false animal repair firm, a "special" and a "chickenhead": "he had been a special now for over a year, and not merely in regard to the distorted genes which he carried. Worse still, he had failed to pass the minimum mental faculties test, which made him in popular parlance a chickenhead."[14] In other words, a character who had "dropped out of history . . . [who had] ceased, in effect, to be part of mankind."[15] J. R. Isidore is the character of the novel who most acutely embodies the condition of the whole universe.[16] He lives alone in "a giant, decaying building which had once housed thousands,"[17] a "dust stricken,"[18] "kipple-ized" building.[19] This latest qualification is Isidore's, for, chickenhead or not, he reflexively embodies junk, that is, his discourse articulates the very metaphysics of the world he inhabits. He worries a lot about *Kipple:*

> Kipple is useless objects, like junk mail or match folders after you use the last match or gum wrappers or yesterday's homeopape. When nobody's around, kipple reproduces itself. For instance, if you go to bed leaving any kipple around your apartment, when you wake up the next morning there's twice as much of it. It always gets more and more . . . the First Law of Kipple . . . "Kipple drives out nonkipple" . . . (one) can roll the kipple-factor back . . . No one can win against kipple, except temporarily and maybe in one spot, like in my apartment

> I've sort of created a stasis between the pressure of kipple and nonkipple, for the time being. But eventually I'll die or go away, and then the kipple will take over. It's a universal principle operating throughout the universe; the entire universe is moving towards a final state of total, absolute kippleization.[20]

In fact, the word *kipple* is an original part of PKD's lexicon, his own word for junk, the materialization of the entropic process.[21] The origin of the word is quite telling because PKD created it, or rather recycled it, from maybe the top-selling postcard of all times, a drawing by Donald McGill (1875–1962) presenting a courting scene with the following caption:

> HE: "Do you like Kipling?"
> SHE: "I don't know, you naughty boy, I've never kippled."

There is thus, ironically, quite a sexual reference buried in the origin of the word, a reference that clearly indicates that PKD connected entropy and generation, a reference that made of Isidore the perfect embodiment of disjunktion: Isidore is a *junk lifer,* condemned to perpetually live among junk. PKD's iron jail is made of junk, and junks its prisoners. In their retrofitting, Scott and his screenplay writers got it, and more. They kept the environment, and made Isidore even more of a junk lifer. J. R. Isidore, PKD's expert in junk, thus becomes in the movie J. F. Sebastian, this man who lives in "a district of silence and ruin (inside) a ten-storey condo gone to shit," and quite appropriately, a geneticist, a designer of the newest artificial form of life, the replicant. And, more accurately, of Nexus-6 replicants. But his contribution to their design is not only a result of his professional skills, and, as he claims in his dialogue with two replicants, there is "some of him in them." In the movie, Sebastian/Isidore cannot emigrate because of a glandular condition ("Methuselah Syndrome"), which causes premature and accelerated aging. Stephen Nottingham noted that "Sebastian's condition also functions as part of the film's elaborate mirroring structure."[22] In the script, Pris, a female replicant, remarks that Sebastian's "accelerated decrepitude" is similar to their problem of a limited life span.[23] So, when Sebastian tells Roy and Pris that there's some of him in them, he is probably (also) referring to their being designed with his defective gene: designed obsolescence! Now, the replicants' limited life span (four years) is both the nexus of the plot and their only alleged inferiority when compared to human beings: Roy Batty and his fellow insurged replicants rebel in order to come back to Earth, confront their creator, and, eventually, obtain a longer lifetime.[24] The replicants' limit, which makes them subhuman as well as a stunning variation on the theme of the overhuman,[25] comes from a "special" who happens to be a "genetic engineer," maybe the most apt characterization of the professional of the future (the knowledge worker of capitalism of the fourth kind): the junk lifer as the origin of another kind of junk life. Could it still be a surprise, then, that, in the novel, Isidore is this character who realizes that replicants are not *actually alive?*

There is yet another level where Ridley Scott's retrofitting leads back to the same genealogy, before van Vogt and his "mysterious chaotic quality." This is the level of the language used in the movie, and especially the names given to things and people. Scott Bukatman aptly noted that

> Scott, revealing an awareness of the textures of science fiction, had been toying with the role of language in his strange new world. He wanted to find new names for the protagonist's profession as well as his targets—detective, bounty hunter and androids were overly familiar terms, no longer evocative enough. Fancher, rummaging through his library, found William Burroughs's *Blade Runner: A Movie,* which was a reworking of an Alan E. Nourse novel about smugglers of medical supplies ("blade runner" also sounds a lot like "bounty hunter", Deckard's profession in the novel). The rights to the title were purchased from Burroughs and Nourse.[26]

I find it highly symptomatic that the very title of the movie actually comes from the work of William Burroughs. And maybe even more symptomatic that this work was itself an adaptation, a screenplay based on an earlier novel. *Blade Runner: A Movie* was never shot, but more than his mere title, some critical aspects of its universe also found their way into Scott's *Blade Runner:*

> Though Ridley Scott adopted the title rather than the story of Burroughs' Blade Runner, the movie's cast of renegade androids and its bosky, evocative ambiance—Piranesian architectural and human ruins outscaled by the monolithic "Mayan pyramids" of the corporate future—belong to Burroughs' fictional world, as does the device of Scott's Alien, i.e., a parasite that eventually consumes and assimilates its host organism.[27]

In Burroughs's screenplay adaptation of Nourse's book, a blade runner is an underground trafficker in medical equipment. The action is set in New York in 2014, a city that "has less a look of having been rebuilt than resettled," and the general ambiance is also built around decay and debris, "derelict skyscrapers and public transport."[28]

The film is about "overpopulation and the growth of vast service bureaucracies" but it is also about "America," "cancer," "the future of medicine and the future of man." It tells yet another form of auto-geddon, political genocide not through wars and camps, but rather by administrative attrition, bureaucratic eugenics implemented through cessation of service for all kinds of minorities. In this world, anyone with a genetically transferable disease must be sterilized, and belonging to a "minority" is considered to be such a "genetic disease." Ethnic or sexual minorities, addicts and psychopaths, are Burroughs's *Blade Runner* future figures for junk life. Their fate is resumed in a classroom lecture of "a shy, pot-smoking professor of bio-mathematics," Dr. Heinz: "the medical miracles of the twentieth century, by destroying natural immunity, results in more illness rather than less . . . Where can this proliferation of recessive genes end? Einstein writes M into E on blackboard. Heinz writes formula on blackboard. Hiroshima. The Health Act Amendment."[29]

Here too redemption comes from the underground, that is, the junked world, not in the form of the android, but rather *in the form of a virus.* Virus B-23, "a virus of biological mutation which restores humanity to pristine health." The cancer epidemic is stopped by virus B-23. "The essence of cancer is repetition—a cell repeating itself like an old joke. I'm a liver a liver a liver . . . Yes an old joke with a halflife of five hundred thousand years."[30] In this perspective, junk life can be adequately described metaphorically as infected life. Junk life appears as a pathological condition, resulting from a contamination. But there is yet another crucial characterization that stems directly from Ridley Scott's *Blade Runner,* and one that was originally developed in PKD's novel: junk life as slave.

From the opening crawler indeed, replicants are presented as "slave labor." Androids, and thus replicants, are the ultimate representation of artificial life, the merging of cybernetic circuits and organic life (the infamous cyborg). As such, they carry the representations ascribed to machines since the dawn of the mechanical age, and especially as "perfect" replacement of human labor, that is, slaves. In fact, the name chosen by PKD to call the ultimate generation of replicants, the "more human than human" Nexus-6, happens to be highly evocative of their function, but also, through its etymology, of yet another resonance with Burroughs's universe, the universe of addiction. Addicts and replicants thus share the same kind of cultural presence since the time of Roman law (cf. chapter 5). In PKD's Gnostic worldview, these are names that can be given by extension to the human person still captive of the iron jail of reality, somebody who needs to be awakened to find redemption. There is numerous evidence of this in his work. We have seen that it is the case in *The Three Stigmata of Palmer Eldritch,* where the alternate realities are "caused" by the consumption of drugs (Can-D and Chew-Z). It is also the case in *Now Wait for Last Year,* where the drug JJ-180 renders the addict "less than human," like "the lizards of the Jurassic Period . . . Creatures with almost no mentalities; just reflex machines acting out the externals of living, going through the motions but not actually there."

In Ridley Scott's *Blade Runner,* Burroughs's perspective on addiction and PKD's Gnostic worldview concur to create a powerful subtext that eventually helps understand its overall visual and scripted presence of junk. Nexus-6, as well as J. F. Sebastian, are thus the archetypical embodiments of junk life, disaffected subjects.

L **Hypervirus**

At the dawn of capitalism's fourth phase, the hypervirus awoke, poisonous parasite, undead, ubiquitous, and omnipotent. At the beginning of the 1980s, the logistic curve of the hypervirus (aka the "virus" virus) passed its first critical point (i.e., second-order inflexion). First (discursive) entity materializing the cybernetic convergence of carbon and silicon, it infected computers and humans alike at unprecedented levels. From then on, explosive diffusion in the "postmodern culture" started, to the point that it eventually plateaued at near saturation, redefining the said culture as a viral ecology. *Room for one more inside, Sir,* as Burroughs used to say. Or, following his postmodern master equation: LANGUAGE = VIRUS = INFORMATIONAL PARASITE.

The hypervirus rules our times like an indifferent despot (it practices the *liberal* indifference). It is the ultimate boot sector parasite of our undead culture. Theoretized, from Derrida to Foucault (who died of it), Baudrillard, and Deleuze, the virus is the master trope of "postmodern culture" (whatever that is).[31] Let us sketch rapidly the progression of the pandemics.[32]

In his Cut-Ups trilogy of the first half of the 1960s (*The Soft Machine, The Ticket that Exploded,* and *Nova Express*), William Burroughs experimented with the stuff of words. In the early 1970s, at the same time Susumu Ohno was coining the *junk DNA* expression—could it be by mere anecdotal synchronicity?—he eventually synthesized the experiment in one fundamental thesis: language (and especially written language) as virus.[33] Approximately at the same time, the "computer virus" appeared in science-fiction literature. William S. Burroughs is *patient 0* of the hypervirus, the original vector. It is an ironic corollary of his own thesis that the hypervirus was first detected in his writings. In *The Electronic Revolution,* he wrote:

> I have frequently spoken of word and image as viruses or as acting as viruses, and this is not an allegorical comparison. It will be seen that the falsifications of syllabic western languages are in point of fact actual virus mechanisms. The IS of identity the purpose of a virus is to SURVIVE. To survive at any expense to

> the host invaded. To be an animal, to be a body. To be an animal body that the virus can invade. To be animals, to be bodies. To be more animal bodies, so that the virus can move from one body to another. To stay present as an animal body, to stay absent as antibody or resistance to the body invasion.[34]

One is reminded here of Deleuze's overman, extended. The virus, more efficient than overman, is not only in charge of the animals, but actually *is* the animals (more on this convergence later). This use of the verb *to be* was, of course, highly problematic for Burroughs, to the point that it is quite accurate to consider him the detective-doctor of the antiviral fight.[35] For him, the principals of this fight began with a reform of language itself, in the tradition of Count Alfred Korzybski's non-Aristotelian semantics, whose seminar he attended in 1938. The first enemy in language, Burroughs thought (after Korzybski), was the "IS of identity": "The word BE in the English language contains, as a virus contains, its precoded message of damage, the categorial imperative of permanent condition."[36] He proposed instead in his reform of language a pictorial (iconic) language where silence is an option. Silence, understood here as the first step in the dissolution of the modern subject (i.e., the egotistical subject,[37] from Descartes on).

At the same time, more or less, another philosophical project mirrored that of Burroughs: between *Of Grammatology* (1967) and *The Dissemination* (1972), Jacques Derrida started a philosophical enterprise attempting to introduce the Other in the I, a redefinition of the subject. Eventually, this "introduction" became "infection," and the Other was radically recast as the virus. Like Burroughs, Derrida first found traces of the process in writing itself:

> The absolute alterity of writing might nevertheless affect living speech, from the outside, within its inside: alter it [for the worse] . . . It is the strange essence of the supplement not to have essentiality: it may always not have taken place. Moreover, literally, it has never taken place: it is never present, here and now. If it were, it would not be what it is, a supplement, taking and keeping the place of the other . . . Less than nothing and yet, to judge by its effects, much more than nothing. The supplement is neither a presence nor an absence. No ontology can think its operation.[38]

No ontology, really? What about a viral ontology, then? Presence/absence, this spatial erasure done, are we left with no-thing? Could we create, following Korzybski and his students, a non-Aristotelian ontology? An ontology of the immaterial supplement . . . Derrida later realized, when the time was ripe, that all he had done since he began writing was "dominated by the thought of a virus, what could be called a parasitology, a virology."[39]

In 1976, Richard Dawkins overextended his selfish gene concept (cf. chapter 1), from the get-go, to ideas: (re)birth of the meme, the other replicator. A nineteenth-century image contemporary to the Darwinian synthesis, *the contagion of ideas,* made possible by a renewed vocabulary: "when you plant a fertile meme in my mind you literally paralyze my brain, turning it into a vehicle for the meme's propagation in just the way that a virus may parasitize the genetic mechanism of a host cell," he wrote.[40] The virus indeed appears as the excluded third that makes the analogy between gene and meme possible.

Dawkins would later make the point even clearer, by referring to certain memes (religious ones) as *mind viruses* (1993), and so opening the door to countless (ab)uses of the metaphor. That same year, the final critical point (second second-order inflexion point) was passed, diffusion was now bound to saturation: the hypervirus was now *In Utero,* as Nirvana sang it (in a Bataillan way), in a song titled "Milk It":

I am my own parasite
I don't need a host to live . . .
I own my own pet virus

But I am getting ahead of myself. Let us come back the false heavens of chronology to describe the epidemics of the timeless entity.

In 1981, Elk Cloner, the first computer virus in the wild (i.e., affecting PCs), was documented, even if every early hacker will tell you that there were programs analog to what we now call "viruses" in the late 1960s or early 1970s.[41] Elk Cloner predated the experimental work that "officially" defined computer viruses and spread on Apple II.[42] When infected, the monitor of the computer displayed the following rhyme:

It will get on all your disks
It will infiltrate your chips
Yes, it's Cloner!
It will stick to you like glue
It will modify ram too
Send in the Cloner!

In 1982, the first global epidemics of the fourth phase officially started: the name AIDS, for Acquired Immune Deficiency Syndrome, was coined in August of that year. AIDS would soon become the syndrome of choice to synthesize and metaphorize the "postmodern condition." In this, it eventually appeared as the final term in a series of diseases playing this part in our culture: plague-tuberculosis-cancer-AIDS. *Room for one more inside, Sir.*

This sequence corresponds term for term to the sequence of the four phases of capitalism: plague is the archaic and thus the archetypical disease (Girard); tuberculosis is the plague of the capitalism of the second kind and its motorized machine, and cancer the disease of the societies of control:

> Early capitalism assumes the necessity of regulated spending, saving, accounting, discipline—an economy that depends on the rational limitation of desire. TB is described in images that sum the negative behavior of nineteenth-century *homo economicus:* consumption; wasting; squandering of vitality. Advanced capitalism requires expansion, speculation, the creation of new needs (the problem of satisfaction and dissatisfaction); buying on credit; mobility—an economy that depends on the irrational indulgence of desire. Cancer is described in images that sum up the negative behavior of twentieth-century *homo economicus:* abnormal growth; repression of energy, that is, refusal to consume or spend.[43]

In this quote, Susan Sontag relates both diseases to an economy of desire. There is a profound resonance here with René Girard's notion of mimetic desire,[44] a resonance that also evokes Richard Dawkins's recycling of the nineteenth-century sociobiologies of imitation.[45] Both, again, were products of the same period, the second oil crisis of international capitalism in the mid-1970s.

In the viral ontology of the capitalism of the fourth kind, the undifferentiating crime is ascribed to the radical Other that is the virus. Metaphorically speaking, the Other then *becomes* a virus. Derrida *is* a virus, concludes the Web author, quoting the same bit where Derrida says is that *all he has done . . . is dominated by the thought of a virus.* The unbearable feedback of the becoming virus . . . "Berlusconi is a retrovirus," writes Lorenzo Miglioli.[46] George W. Bush is a virus, Saddam Hussein is a virus, and bin Laden is a virus. *Room for one more inside, Sir.*

In 1983, on November 3, the first "official" computer virus was conceived of as an experiment to be presented at a weekly seminar on computer security. Frederic Cohen first introduced the concept in this seminar, and his PhD supervisor, Len Adleman, proposed the name "virus." In his presentation, Cohen defined a computer virus as "a computer program that can affect other computer programs by modifying them in such a way as to include a (possibly evolved) copy of itself," a definition he would stick to in his subsequent paper,[47] and one that would become the official definition of a "computer virus." He demonstrated such an "infection" within a Unix directory-listing utility, and proved that identifying and isolating computer viruses was a noncomputable problem. This later result, maybe the most crucial point in Cohen's work, meant that fighting the infection is therefore impossible to achieve using an algorithm, and one is left with the same aporia that philosophers had diagnosed.

According to Cohen, the first use of the term "virus" to refer to unwanted computer code occurred in 1972 in a science-fiction novel, *When Harley Was One,* by David Gerrold. In an interview, Len Adleman concurred with Cohen: "The term 'computer virus' existed in science fiction well before Fred Cohen and I came along. Several authors actually used that term in science fiction prior to 1983. I don't recall ever having seen it, perhaps it was just a term whose time had come. So I did not invent the term. I just named what we now consider computer viruses 'computer viruses.'"[48] Yes, indeed, it was a term whose time had come! And the convergence was not fortuitous, since Alderman later went on to propose a new mathematical description of AIDS.[49]

In 1986, the diffusion curve of the hypervirus passed its first-order inflexion point, and it thus became mainstream. That year, two Pakistani programmers replaced the executable code in the boot sector of a floppy disk with their own viral code designed to infect each 360kb floppy accessed on any drive. Their "Brain virus" (infected floppies had "© Brain" for a volume label) became the first recorded virus to infect PCs running MS-DOS. It was also the first "stealth" virus, meaning it tried to hide itself from detection. If a computer user tried to view the infected space on the disk, Brain would display the original, uninfected boot sector. Yep, Burroughs and Derrida could have anticipated a dialectics of presence/absence . . . That same year, the performance artist Laurie Anderson turned William Burroughs's original insight mainstream.[50] One year later, AIDS turned mainstream too, thanks to a "hit" from Prince:

> Oh yeah
> In France a skinny man
> Died of a big disease with a little name
> By chance his girlfriend came across a needle
> And soon she did the same.[51]

As in this song, the syndrome, however, was still restricted to certain stigmatized groups (homosexuals, junkies, etc.). At first, indeed, the syndrome was dubbed "the gay cancer." Contrary to the other three diseases associated with prior phases of capitalism, it is highly significant that the main mode of AIDS transmission was by sexual contact. In 1988, Susan Sontag had already understood this, when she followed her original essay on cancer and "Illness and Metaphor" with an update focusing on AIDS. She wrote: "The sexual transmission of this illness, considered by most people as a calamity one brings on oneself, is judged more harshly than other means—especially since AIDS is understood as a disease not only of sexual excess but of perversity."[52] It is quite well expressed in this other 1987 song by the Pet Shop Boys:

> Now it almost seems impossible
> We've drunk too much, and woke up everyone
> I may be wrong, I thought we said
> It couldn't happen here . . .
> It contradicts your battle-scars
> Still healed, so far.[53]

And the Boss himself concurred, some years later, when the time was ripe for a cinematographic representation of an AIDS patient as a white lawyer (Tom Hanks got the part).[54] The year before (1992), 1,300 computer viruses were recorded, an increase of 420 percent from December 1990. By November 1990, one new virus was discovered each week. Today, between ten and fifteen new viruses appear every day. In fact, from December 1998 to October 1999, the total virus count jumped from 20,500 to 42,000. I guess that one will soon stop counting; we've got spyware (aka junkware, *stricto sensu*) now, and that too was kind of anticipated by Burroughs:

> It is worth noting that if a virus were to attain a state of wholly benign equilibrium with its host cell it is unlikely that its presence would be readily detected OR THAT IT WOULD NECESSARILY BE RECOGNIZED AS A VIRUS. I suggest that the word is just such a virus.[55]

It is worth noting indeed that the ambiguity about the hypervirus is essential to its functioning as the master trope of the postmodern condition. If AIDS is the syndrome of choice to concretize the hypervirus in the postmodern culture, one should remark that, strictly speaking, and contrary to the three diseases associated with prior kinds of capitalism, AIDS is no disease, but a *syndrome*. AIDS is the name of a medical condition associated with a wide spectrum of diseases that are usually assumed to be the consequences of the HIV infection. But this very point is still the subject of a controversy as I write these lines. If most of the medical and scientific community accepts today that AIDS results from the HIV infection, this is not exactly a proven fact, and some say (e.g., the group of Perth; Kerry Mullis, 1993 Nobel Prize for chemistry) that it is only still a hypothesis—and a bad one at that.[56] In this sense, AIDS/HIV is quite a stealth virus, as computer scientists would say. Rather than a mere epiphenomenon of big science, I consider this point as a crucial characteristic of the hypervirus.

The resonance between Burroughs's notion of the word as virus and Deleuze and Guattari's concept of the rhizome is far from fortuitous. In fact, Deleuze and Guattari took the virus as the one example (apart from the wasp and the orchid) to explain deterritoralization and the rhizome:

> Under certain conditions, a virus can connect to germ cells and transmit itself as the cellular gene of a complex species; moreover, it can take flight, move into the cells of an entirely different species, but not without bringing with it "genetic information" from the first host . . . We form a rhizome with our viruses, or rather our viruses cause us to form a rhizome with other animals . . . Always look for the molecular, or even submolecular, particle with which we are allied. We evolve and die more from our polymorphous and rhizomatic flux than from hereditary diseases, or diseases that have their own line of descent. The rhizome is an anti-genealogy.[57]

Mark Hansen has discussed this convergence at length, and he defended the thesis that Deleuze and Guattari took Burroughs's virology beyond its cybernetic limits—and in particular the limits of Claude Shannon's theory of information, which, along with Korzybski and Wilhelm Reich, had provided its main inspiration:

> Whereas Burroughs correlates the virus (language) with a host (the body, affective life) without which it could not survive (being itself only quasi-alive), D+G decorporealize the virus, decoupling it from any host and granting it an autonomy and agency to produce new connections, new bodies, and in fact, what D+G felicitously call "nonorganic life." By wresting the virus from its molar correlation with a host body and setting it free within the domain of the molecular (the plane of immanence), D+G are thus able to depathologize the virus.[58]

The status of the virus vis-à-vis the living is still problematic today. As I first wrote these lines, the December 2004 edition of *Scientific American* was standing out amid the jumble of my desk, with this simple question on its cover: "Are viruses alive?" On page 105, the monthly publication reproduced an affirmation dating from 1962 by the French laureate of the Nobel Prize in physiology or medicine, André Lwoff: "Whether or not viruses should be regarded as organisms is a matter of taste."[59] So little has changed since the 1960s regarding this question; however, a profound change in perspective has taken place. While humankind is experiencing its first alleged retroviral pandemic (AIDS), the virus becomes the site of a fundamental scientific controversy: parasite or symbiont (or, alternatively, both)?

The works of Lynn Margulis and her son Dorion Sagan are often cited as the central reference for the second position.[60] Hansen is no exception to the rule: in their work, he detects a resonance with the work of Gilbert Simondon, whom he says provides the means to reach beyond the virologies of Burroughs and of Deleuze and Guattari.[61] Similarly, but this time following Deleuze and Guattari all the way, Luciana Parisi situates her recent work in the perspective of Lynn Margulis's theory of endosymbiosis, as the "constitutive process of the abstract machine of sex or abstract sex . . . a rhizomatic conception of the evolution of sex proceeding by contagion rather than filiation."[62]

No need to continue to multiply the examples: you will have understood that an entire Deleuzio-Guattarian exegesis—by operationalizing contemporary questions on theories of evolution—actualizes this positive convergence with the works of Margulis and Sagan. The result is the virus's depathologization, as Hansen puts it, but, even more, its redefinition as a *driver of evolution*, by genome or partial-genome fusion-acquisition. This was never clearer than during the unveiling of the first draft of the human genome in 2001, when David Baltimore, who "discovered" reverse transcriptase, qualified the genome as "a sea of reverse-transcribed DNA, with a small admixture of genes."[63]

If Bruno Latour once described the biochemist as the "last of the rogue capitalists,"[64] one must see in the virus, often the "object" of the biochemist's studies, the first of the genetic capitalists, and in the genomic viral fusion-acquisition, the essential principle[65] of genetic capitalism, this fourth phase of creative destruction (according to Schumpeter's expression). Thus, as Keith Ansell Pearson so aptly sums it up: "Today the life of the great empires has assumed a retroviral form, fragmented and peripheral, genetically infecting their wastes and by-products."[66] Mirroring Kurt Cobain's insight, his injunction might be "Embrace your viruses!" or, even more, "Embrace yourself as a virus." Steven Shaviro, in his "Two Lessons from Burroughs," proposed such a "biological approach to postmodernism," and offered violent viral replications and insect strategies such as swarming as models.[67] In a Deleuzian fashion, he suggested that we find out about the Other by becoming other, by posing "the question of radical otherness in biological terms, instead of epistemological ones . . . resolving such a problem would involve the transfer, not of minds, but of DNA."[68] This transfer today is called "lateral," what Deleuze and Guattari called "aparallel."[69]

What used to be seen as a problem, or even a stigma, is now portrayed as a path to freedom, in a highly paradoxical statement strongly reminiscent of Philip K. Dick's Gnostic theodicy. For the ambivalence, of course, remains. As I first wrote these lines, my native country was agitated by the aftershocks of the declarations of a comedian who has quite simply actualized the cultural ambivalence of the hypervirus total diffusion in an aphorism equating Zionism with "the AIDS of Judaism."

I was reminded then of the famous characterization of my own generation by Louis Pawels, in an editorial for *Le Figaro Magazine* in December 1986, as "suffering from mental AIDS." As we were demonstrating in the streets against one more reform of the educational system, Pawels wrote that we, "the children of stupid rock, the pupils of pedagogical vulgarity," had "lost our natural immunities."[70] Those viruses that were supposed to infect us, were, of course, "mind viruses," as Dawkins would have put it. By the time Pawels passed his judgment on my generation, AIDS was definitely going mainstream, and "low culture" (i.e., rock 'n' roll and vulgar pedagogy) had rejoined sex and drugs to complete the list of the symptoms of the hyperviral infection. Most of us shrugged, laughing, and passed the joint.

Today, such metaphoric uses of AIDS are so common that nobody seems to notice them anymore. Today, a little googling brings back the following instances from the Web: AIDS as a metaphor for violence, apathy, fear, loneliness, colonialism, globalization, pollution, ecological collapse, homosexuality, the opposing basketball team (!), the corruption and betrayal of the masses, chronic illness, the social and political deterioration of a fictional country, the general loss of moral standards, the conflicts tearing at American society at the turn of the millennium, the American condition, inequities, social decay, or merely "how the world works." *Room for one more inside, Sir.*

But there is yet one more crucial level where today's troubled times are understood by the AIDS metaphor: terrorism as a consequence of "metaphysical AIDS." This one we owe . . . to Jacques Derrida.[71] In an interview with Giovanna Borradori that took place in the wake of the 9/11 "event," he developed this thesis: terrorism as the latest symptom of (Occidental) suicidal autoimmunity.[72] Borradori noted quite interestingly that Derrida began his reflection on the mechanism of autoimmunization during the winter of 1994, "in connection with a study of the concept of religion, which frames his discussion of religious fundamentalism and its role in global terrorism."[73] And Derrida agrees, pointing back to a text written during that time: "In analyzing 'this *terrifying* but inescapable logic of the *autoimmunity of the unscathed* that will always associate Science and Religion,' I there proposed to extend to life in general the figure of an autoimmunity whose meaning or origin first seemed to be limited to so-called natural life or to life pure and simple, to what is believed to be the purely 'zoological,' 'biological," or 'genetic.'"[74]

Now, remember: 1994 is the same year when Derrida realized that all his prior work from *On Grammatology* on could be reinterpreted as a kind of virology. In 1994, Sadie Plant and Nick Land reached the logical conclusion that has served as a premise for *Junkware,* linking DNA, Baudrillard's "prophet" of the metaphysics of code, with a cyberpositive capitalist culture marked by immuno-politics, under the sign of "the virus," that is, junk code:

> Capitalism is not a human invention, but a viral contagion, replicated cyber-positively across post-human space . . . Viruses are tangible transmission, although you only know about them when they communicate with you: messages from Global Viro-Control. Viruses reprogram organisms, including bacteria, and even if schizophrenia is not yet virally programmed it will be in the future . . . The linear pathway from DNA to RNA is the fundamental tenet of security genetics. The genotype copies God by initiating a causal process without feedback. But this is merely superstition, subverted by retroviruses. Viral reverse transcription closes the circuit, coding DNA with RNA, switching the cybernetics to positive.[75]

Bless them for such an insight! They also wrote about Burroughs, the city, Yage, LSD and drugs, Gibson's ice, and immuno-politics: another junked table of contents. But they did not talk about terrorism, at least not directly. Maybe indirectly, they concluded: "only the enemies of immuno-identity populate the future." In 1988, Susan Sontag had also written about the same process: "In the description of AIDS the enemy is what causes the disease, an infectious agent that comes form the outside . . . Next the invader takes up permanent residence, by a form of alien takeover familiar in science-fiction narratives. The body's own cells *become* the invader . . . What makes the viral assault so *terrifying* is that contamination, and there *vulnerability,* is understood as permanent."[76]

Why, then, this elision of the virus in Derrida's account of terrorism, why this strange feeling that if terrorism amounts to suicide, it is a spontaneous autophenomenon, with no external agent? In his first moment or first autoimmunity, Derrida provides an answer. The aggression is coming from the inside because it comes from "forces that are apparently without any force of their own but that are able to find the means, through ruse and the implementation of *high-tech* knowledge to get hold of an American weapon in an American airport."[77] But that too could be said of viruses. More important, Derrida adds, "let us not forget that the United States had in effect paved the way for and consolidated the forces of the 'adversary' by training people like 'bin Laden' . . . and by first of all creating the politico-military circumstances that would favor their emergence."[78] As in a kind of engineered virus, then? No, answers Derrida, *doubly suicidal.*

If Derrida does not see the stigmata of the hypervirus in 9/11, it might also be because it would amount to the repetition of Jean Baudrillard's thesis. Earlier than 9/11, even earlier than the time Derrida understood that his work was producing a kind of virology, Baudrillard started to understand terrorism as one symptom of the hyperviral infection. Like Derrida, he understood it as the result of a suicidal drive: "The terrorist hypothesis is that the system itself suicides in response to the multiple challenges of death and suicide."[79] But unlike Derrida, he resorted to a viral explanation, even if it did not take the face of an "external adversary":

> Terrorism, like viruses, is everywhere . . . with terrorism—and its viral structure—, as if every domination apparatus were creating its own antibody [anti-dispositif], the chemistry of its own disappearance; against this almost automatic reversal of its own puissance, the system is powerless. And terrorism is the shockwave of this silent reversal.[80]

This was exactly my point: the very core of the culture that fights the hypervirus—postmodern theoreticians included—is infected by it. Terminally. Terrorism is but one—albeit a crucial one—symptom of the infection. It reflects the vital (and morbid) condition of postmodernity, setting the stage for the fourth phase of capitalism. Both source of pain and suffering and maybe the only sign of a future to come, a junk future. Could this future only be death, as *patient 0* seemed to have concluded?

> "Fight tuberculosis, folks."
> Christmas Eve, an old junkie selling Christmas seals on North Park Street.
> The "Priest," they called him. "Fight tuberculosis, folks." . . .
> Then it hit him like heavy silent snow.
> All the gray junk yesterdays.
> He sat there received the immaculate fix.
> And since he was himself a priest,
> there was no need to call one.[81]

Junk is yet another name of the hypervirus: in Burroughs's mind, virus and junk are connected through the power of the image, another excluded third. From the awakening of the hypervirus in *Nova Express*, Burroughs had realized that "junk is concentrated image" and that "the image material was not dead matter, but exhibited the same life cycle as the virus."[82] *Now I offer this: a virus is essentially junk code, and our hyperviral culture is indeed a junk culture.* The hypervirus is junkware, transcending hard-, soft-, and wetware, neither dead nor alive, past good and evil, the metaphor of choice for a time of disarray.

K **Junkyard Terror**

I grew up taking hikes in the woods where my grandmother used to live and where my father was born, nearby Le Chemin des Dames, quite a romantic name for one the worst battlefields of World War I. There, I used to pick mushrooms and often would find more evidence of the killings—bullet-punched helmets, discarded boots, or even whole shells—than of my precious morels. There, I guess, I decided to become an agronomist, out of idealism, and to get back to the countryside that my parents, like their whole generation, had deserted for the Paris suburbs.

I have known terrorism since I was a young adult. In my late teens and early twenties, the streets of Paris, whcrc I uscd to live, were frequently the theater of terrorist operations. Bars, synagogues, train or metro stations were blown up once in a while. Like Baudrillard, and like millions of soccer fans, I witnessed European stadiums erupt in barbaric violence. For a while, terror meant taking the subway: a new trend in violence had appeared, when some pricks made a habit of pushing other passengers under the incoming train—quite a common mode of suicide at that time, recycled for "fun."

Later, I came to realize that political modernity had started with terror, with Saint Just's head rolling down in a basket, if I have to pick a starting event. Modernity started with the instauration of state violence (against its own people) for revolutionary purpose (for the alleged good of the same people). Since then, and for two centuries, no democratic state could avoid this fatality.

During the twentieth century, no matter what revisionists might argue, mankind passed a threshold in its exercise of violence. Pogroms, wars, gulags, camps, and all sorts of *dispositifs de mort* actualized violence in unprecedented ways: in Europe, first, we industrialized it. It is not a question of numbers; past a certain quantitative level (a threshold), change becomes qualitative. Somewhere between the earliest camps, during the American Civil War, and Auschwitz, we human beings moved from archaic (agrarian-despotic) forms of violence to industrialized violence. Mass-produced and mass-delivered. In this respect, modernity is defined by state mass violence, that is, legal violence: the camp, this paradigm of modernity, albeit often in a state of exception, still belongs inside the boundaries of legality (Agamben).

At first, like so many European migrant workers to America, I thought that I had left terrorism behind me when I left France. I had, indeed, quite an idealized America in mind.

Sometime between the events of Oklahoma City (1995) and the World Trade Center (take 2, 2001, for take 1 took place in 1993), I changed my mind.

In between, a peculiar character got my attention: dubbed the Unabomber when his arrest made a "pop icon" out of him, he was first nicknamed the "junkyard bomber" by the FBI team that ran after him for eighteen years before it eventually caught him.[83]

The junkyard bomber practiced singular terrorism out of recycled parts. He took a meticulous pride in his craft. He picked individuals and sent them small-area-coverage, passive, handmade bombs, by mail most of the time.[84] The U.S. postal system was thus a part of his killing *dispositif.* Recycling was his business; he redefined junk mail (for the worst). Five years after his fall and eventual incarceration for life, some other terrorists followed his trail, recycled with horrifying success the American airlines, and junked three planes in the heart of the Empire, killing more than three thousand people in the process. The junkyard bomber was a trendsetter.[85] History was back, and terrorism definitely became the latest version of the hyperviral infection, the ultimate Occidental virus (or autoimmune syndrome, as Derrida had it).

The horrendous odyssey of the junkyard bomber parallels accurately the "progress" of the hyperviral infection. He started thinking about his terrorist project in the early 1970s, at the time of the *Electronic Revolution,* a time when terrorism was making a remarkable comeback in the United States. From the second coming of the KKK in the South to the California Rangers, the Order, the Eco-Raiders, and so on, rebel America (re)discovered a new mode of action. Too "chicken" at first (in his own words), the junkyard bomber would eventually take a few years to find the guts (and despair) to write in his journal "I think that perhaps I could now kill someone."[86] Centered on the hyperviral first critical threshold, he went on his first killing spree (1978–82: seven bombs); a second one would follow around the second critical threshold (1985–87: five bombs) and a third and final around at the time of last critical threshold (1993–95: four bombs).

His devices were Weapons of Singular Destruction (WSD). He built them out of junk—cigar boxes, copper pipes, lamp cords, sink traps, furniture parts, old nails and screws, and even human hair. He built a pistol from scrap in 1980, and intended to use it as a "murder weapon."[87] The junkyard bomber was a tinkerer, a "cool-headed" experimenter who conceived his devices and their uses in a very rational way, with ample notes and pictures in his logbook, pretty much as a scientist would. The style of his writings was as devoid as possible of any emotion, dis-affected. They were, plainly, reports of an ongoing experiment. An experiment in killing.

Indeed, Theodore Kaczynski, aka junkyard bomber, aka unabomber, was one of us professors when the venerable institution that is named "The

University" started to fall into ruins. He completed his undergraduate degree in mathematics at Harvard, graduating in June 1962, at the age of twenty. He began his first year of graduate study at the University of Michigan at Ann Arbor in the fall of 1962. He completed his Masters and PhD by the age of twenty-five. Following graduation, he accepted a position as assistant professor in the math department at the University of California at Berkeley, a position he occupied until June 1969. One of his Harvard classmates called him "one of the last Berkeley professors to wear a jacket and tie at work."[88] During the summer of 1969, he left for the wild, never to occupy an academic position again.

The FBI task force nicknamed him the Unabomber because he picked his victims mostly from the ranks of the university (Un-) or the airlines (-a-): Un-a-bomber. Kaczynski, however, resented this characterization, and made it pretty clear in a letter to the *New York Times:* he was merely "out to get scientists and engineers, especially in critical fields like computers and genetics."[89] Kaczynski's problem was control, "the system's use of behavioral modification techniques to control the individual." FC, his signature, stood for *Freedom Club,* his "manifesto" was a revolutionary outcry for the restoration of individual freedom, which he defined as "participation in the power process."[90] Alston Chase, another Harvard graduate, developed a fascinating thesis in his book about the Unabomber case. According to him, Kaczynski encountered at Harvard a "culture of despair" (Chase's name for nihilism) in the form of the "general education curriculum" that fed him most of his ideas about a "system" gone wrong:

> Kaczynski's Harvard experiences shaped his anger and legitimized his wrath. By graduation, all the elements that would ultimately transform him into the Unabomber were in place: the ideas out of which he would construct a philosophy; the dislike of mathematics and psychology; the unhappiness and alienation. Soon after, too, would come his commitment to killing. Embracing the value-neutral message of positivism—morality was merely emotion—made him feel free to commit murder.[91]

But, as Chase realized, many undergrads at Harvard (and pretty much in every other institution of higher learning in the 1950s and 1960s) were exposed to such ideas—to the point that they had become a *compendium of clichés*[92] by 1970—and yet only one of them became the Unabomber. Among Kaczynski's Harvard *experiences,* however, Chase claims that there was one that put him over the edge: his participation in a psychology experiment led by Professor Henry A. Murray. Titled "Multiform Assessments of Personality

Development among Gifted College Men," it centered on a procedure whose aim, according to Chase, was to "catch the student by surprise, to deceive him, bring him to anger, ridicule his beliefs, and brutalize him."[93] Chase goes on for pages exposing the relationships between Murray and the OSS (the CIA precursor) and his twisted personality. Most of Chase's critics—not surprisingly, many of whom happen to be Harvard men—found this thesis "absurd"[94] or, at the very least, far-fetched.[95] Whatever the case, Chase has a point: Kaczynski's main issue was "behavior modification techniques," of which he felt he had been a *victim* (among other things, including alleged parental abuse). My point here is not to dwell on psychological explanations, but rather to notice that this is the core of Kazcynski's beliefs—the root of his particular breed of evil, delusional or not—and that it organized his action. He focused on university professors, computer and biology scientists, public-relation people, because, in his mind, they were part of the system's organization and use of "behavior modification techniques," as this excerpt of his infamous *manifesto* attests:

> 149. Presumably, research will continue to increase the effectiveness of psychological techniques for controlling human behavior. But we think it is unlikely that psychological techniques alone will be sufficient to adjust human beings to the kind of society that technology is creating. Biological methods probably will have to be used. We have already mentioned the use of drugs in this connection. Neurology may provide other avenues of modifying the human mind. Genetic engineering of human beings is already beginning to occur in the form of "gene therapy," and there is no reason to assume that such methods will not eventually be used to modify those aspects of the body that affect mental functioning.

This is probably why he threatened Phillip Sharp, among other biological or "gene therapy specialists." In what might amount to the "ultimate junk mail," the junkyard bomber sent the following letter, dated April 20, 1995, to the researcher: "Dr. Sharp it would be beneficial to your health to stop your research in genetics. This is a warning from FC. Warren Hoge of the *New York Times* can confirm that this note does come from FC." This was a warning only, and the letter did not explode when Phillip Sharp opened it. At that time, Sharp was head of the biology department at MIT. He had shared the Nobel Prize in medicine in 1993 with Richard J. Roberts "for their discovery of split genes," this "mini revolution in molecular genetics" (*dixit* Francis Crick, see chapter 1). Since Roberts received the same letter, one must conclude that it is because of this singular achievement that the junkyard bomber was after them. This was either puzzling, or trivial, or far-reaching on the part of Kaczynski. It could possibly mean that he went after the Nobel symbol (trivial), or that he targeted research rather than development (puzzling or far-reaching). Sharp, for instance, was already on the board of directors and even chaired the scientific advisory board of Biogen, one of the earliest pioneer companies in biotechnology.[96]

Whatever the case, it seems a tragically ironic turn of events that made of one of the discoverers of biomolecular junk a potential victim of the junkyard bomber. Junk culture eats its own children.

I **Junkspace**

"Intellectual terrorist," "architect of warehouses," "agitator of ideas," and "star of postmodern architecture," Rem Koolhaas evokes qualitative hyperbole. Born in the ruined Rotterdam of 1944, this architect, journalist, writer, entrepreneur, and now professor in the department of architecture at the prestigious Harvard Design School, leaves no one indifferent. He has been called all of the following: cold, indifferent, cynical, mean, kind, provocative, solicitous, master strategist, and lacking technique; in short, he is described as the last of the architects, or the last architect, as suits an era that is seemingly only distinguishable by the deaths it announces, which are celebrated with grand (and often empty) pomp.

Critics do not always know how to label his accomplishments, and they hesitate primarily between "postmodern" and "deconstructivist" (whether as compliment or insult). Some insist on their avant-garde character, noting that "one often has the impression that Koolhaas emerged from poor neighborhoods, offering a glimpse of the shameful world of business so as to distinguish his brand of avant-gardism from the avant-garde dominated by the outdated theory of the 1980s and 1990s."[97] Others, more contentious, see only "neo-modern non-sense" that tries "to take on chaos and use it as a poetic machine," "nothing other than voluntary resignation, a guilty penchant for facility" (Michel Antonietti). In any case, Koolhaas disturbs even his epigones, who, like Fredric Jameson, too often attempt a monolithic reading, à la postmodern; for example: "the point of the exercise is rather to find synonyms, hundreds upon hundreds of theoretical synonyms, hammered one upon the other and fused together into a massive and terrifying vision, each of the 'theories' of the 'postmodern' or the current age becoming metaphorical to the others in a single blinding glimpse into the underside."[98]

Koolhaas, however, indicated what he thought of this: "Post-modernism is fundamentally a return to the past. It was a process of copying originals, copies of copies, imitations of interpretations, all timidly following the past. It not only looted our past, but even more, has deprived us of our present and obliterated our future."[99] Here, I am interested in the architect of the word, the one who has understood that the modern city, such as Italo Calvino's *Clarisse* and *Trude*, is "a *place* and a *name* that identifies it, and that becomes little by little . . . that of its airport."[100] It is the urbanist of the species that interests me here, the one who wonders about the future of humankind at the dawn of this millennium, in relation to his habitat, his modern hell. Koolhaas began by calling Calvino's *Clarisse/Trude* "the generic city," paraphrasing his (implicit)[101] reference as early as the first sentence of the introduction:

> Is the contemporary city like the contemporary airport—"all the same"? Is it possible to theorize this convergence? . . . Convergence is possible only at the price of shedding identity. That is usually seen as a loss. But at the scale at which it occurs, it *must* mean something. What are the advantages of identity, and conversely, what are the advantages of blankness? What if this seemingly accidental—and usually regretted—homogenization were an intentional process, a conscious movement away from difference toward similarity? What if we are witnessing a global liberation movement: "down with character!" What is left after identity is stripped? The generic?[102]

What Calvino might seem to deplore ("the crisis of the too-large city is the reverse of the crisis of nature," he writes),[103] Koolhaas embraces. In short, at first sight, Koolhaas strategically chooses two entries in *Les villes invisibles*. In *inferno*, he cites the last sentences of Calvino's book:

> And Polo said: "The inferno of the living is not something that will be; if there is one, it is what is already here, the inferno where we live every day, that we form by being together. There are two ways to escape suffering it. The first is easy for many: accept the inferno and become such a part of it that you can no longer see it. The second is risky and demands constant vigilance and apprehension: seek and learn to recognize who and what, in the midst of the inferno, are not inferno, then make them endure, give them space."[104]

In his preface written for a later edition (1993, French translation 1996), Calvino emphasizes the fact that even if "almost all the critics stopped on the last sentence in the book [and that] all thought it was the conclusion, the 'moral of the story,' his book was no less constructed as a polyhedron, with conclusions inscribed here and there, along all the way, and some have no less allure as epigram or epigraph than that last one."[105] It is such an epigraph, I think, that Koolhaas chose for his second entry in invisible cities:

> "I have also thought of a model city from which I deduce all the others," Marco answered. "It is a city made only of exceptions, exclusions, incongruities,

> Contradictions. If such a city is the most improbable, by reducing the number of elements, we increase the probability that the city really exists. So I have only to subtract exceptions from my model, and in whatever direction I proceed, I will arrive at one of the cities that, always as an exception, exists. But I cannot force my operation beyond a certain limit: I would achieve cities too probable to be real."[106]

Synthesized, these two entries might mean something like the following: the manner chosen by Koolhaas to not suffer *inferno* is to construct *exceptions* for it. All of his work can be read as research for the "certain limit of the plausible/probable" evoked by Calvino (pardon me, by Marco Polo). As confirmation, I point to the third entry by Calvino's pen in Koolhaas's dictionary, *photography:* "Perhaps the true, total photography, he thought, is a pile of fragments of private images, against the creased background of massacres and coronations."[107] For the generic city completes the modern city, as in the conclusion of the text where Koolhaas invites us to witness a spectacular fiction, "a Hollywood movie about the Bible," more specifically a scene of a riot in a market, without sound, and projected backwards (reminiscent of Debord's proposition: "in a truly inversed world, the real is the moment of the false"):[108]

> The viewer no longer registers only human characters, but begins to note spaces between them. The center empties; the last shadows evacuate the rectangle of the picture frame, probably complaining, but fortunately we don't hear them. Silence is now reinforced by emptiness: the image shows empty stalls, some debris that was trampled underfoot. Relief . . . it's over. It's the story of the city. The city is no longer. We can leave the theatre now.[109]

Fragments, debris, sundry vestiges, the table (of contents) is set for the theorization of *junkspace.* The city is dead, or, more precisely, the city has lived out the destiny of all lives subject to the laws of thermodynamics: modernity won't survive entropy, the late-modern city collapses under its debris, to the point of becoming *unidentifiable.* Such is the conclusion of *S, M, L, XL,* the first twenty years of the Office for Metropolitan Architecture's work (1975–94), and such are the firstfruits of its subsequent work. Koolhaas eventually found *Clarisse/Trude*'s generic name, more than a theory, *junkspace:*

> The built product of modernization is not modern architecture but Junkspace . . . Traffic is Junkspace, from airspace to the underground; the entire highway system is Junkspace, a vast potential utopia clogged by its users . . . Junkspace is post-existential; it makes you uncertain where you are, obscures where you go, undoes where you were . . . Junkspace is political: it depends on the central removal of the critical faculty in the name of comfort and pleasure . . . Junkspace

> is both promiscuous and repressive . . . all Junkspace's prototypes are urban—the Roman Forum, the Metropolis; it is only their reverse-synergy that makes them suburban, simultaneously swollen and shrunk . . . Junkspace is like a womb that organizes the transition of endless quantities of the Real into the virtual.[110]

For Rem Koolhaas, *Junkspace* is the contemporary concept of a public space that is increasingly privatized, made spectacular.[111] The *Harvard Design School Guide to Shopping* begins with the following proposition: "SHOPPING is arguably the last remaining form of public activity," and ends with a text titled "After Private Space." *Junkspace* is neither private nor public, but privatized and spectacular: it is process-space. It is the interstitial space invaded inch by inch by merchandise and the procession of its simulacra. The ¥€$ of the generic city transforms itself progressively, while modernity triumphs, in a planetary mall under the combined action of the three hypostases of the *inferno:* air conditioning, escalator, and glass. Covered by ads, signs, billboards, posters, displays, neon lights; oversaturated by muzak, artificial smells, people. Busy individuals, in search of product, or simply there to kill time, such as those American teenagers for whom this is the gathering place, the contemporary *summer camp.* Maidens licking attractive window displays, finding sustenance at the multiethnic food courts and other fast-food temples, accosted by the echoes of loiterers barely finished with their experience of the latest blockbuster. *Junk way of life.*

The fast-food stand started out along the highway,[112] then invaded the train station, the metro station, the airport . . . tomorrow, the cosmodrome. The blockbuster is first and foremost the name of a weapon of mass destruction during the Second World War. The mall is the ultimate version of the Parisian *arcade,* when that city was the capital of the nineteenth century: "thus the passage is a city, a world in miniature."[113] *Junkspace,* the windup of constructed modernity, is the concentration camp of the consumer. The last text of the collection insists: *junkspace* is also a space of control where the consuming singularity can be permanently located by the traces she leaves behind (credit transactions, loyal customer cards, smart cards, etc.). The space of the encumbered rest of the commercializable debris of modernity, *junkspace* is populated by entities devoted to consumption and detectable by their discrete tracks, pinpointed like data in an omnipresent database. If space has a becoming junk, as Koolhaas proposes, isn't it this interstitial space, this rhizomatic space? Indeed, the first word that Koolhaas borrows from Deleuze and Guattari is *capitalism,* and *junkspace* participates in the global future of capitalism: Roman forum (110), bazaar of Isfahan (1585), Parisian arcades (1815),

London's Crystal Palace (1851), the high-end architecture of Macy's, Marshall Field's, and JCPenney (1902), the Country Club Plaza of Kansas City (1922), Southdale in Minnesota (1956), Las Vegas (2000) . . .

I would now like to conclude with the questions Koolhaas raises at the end of *Junkspace* by considering the future of the human body and species: "Will Junkspace invade the body? Does gene therapy announce a total reengineering according to Junkspace? Is it a repertoire of reconfiguration that facilitates the intromission of a new species into its self-made *Junkbiosphere?*" Indeed, it seems to me that here Koolhaas is outdated by his own intuition, since the junk paradigm has already invaded the human body in the most intimate fiber, memory, and program: its DNA.

As Deleuze and Guattari say, *the link has become personal.* To go one step further, will the person become a link, essentially *junk? Junkspace* is the smooth space of the machine of the fourth kind, the translation in (urban) space of genetic capitalism's generalized equivalence, language as virus and virus as language. Fredric Jameson understood this well when he titled the section of his text dedicated to *junkspace* "Down with the *junkspace* virus": "junkspace becomes a virus that spreads and proliferates throughout the macrocosm . . . Junkspace has been around for some time, at first unrecognized; again, like a virus undetected."[114] This intuition resonates with most of the streams we have followed so far, from PKD and William Burroughs to Ridley Scott, to the point that Koolhaas's junkspace appears to be the conceptual retrofitting of *Blade Runner* cinematographic presence of junk, leading to the actual building of our urban landscapes into junkspaces.

One more loop into the recycling bin, Dr. Koolhaas!

J Future Eves, Artificial Menials, and Capitalist Re-genesis

I extracted half of my title for this section from one of the leading modernist tales of the artificial creature, Villiers de L'Isle-Adam's *Future Eve* (1886). In doing so, I want to start with the female archetype of the android, the "artificial bride," and come back to his masculine counterpart, the artificial servant, later.[115] Villiers de l'Isle-Adam's account is important because he anticipated a later trend of modernity. "Since our gods and hopes are only scientific now," he wrote, "why shouldn't our loves become scientific too? Instead of the forgotten legendary Eve, of the legend despised by Science, I offer you a scientific Eve—only worthy, it seems to me, of these withered viscera that—from a remainder of sentimentalism of which you are the first to laugh—you still call 'your hearts.'" His artificial Eve—named Hadaly for "Ideal"—is the quintessential *sexyborg,* as seen from the masculine trenches of the sex wars.

"Electric Daughter" of the famous Edison, she is everything a man can desire, *plus* a female creature with no desire for men. It is Edison here who plays the part of the Master, artist, magician and scientist, maker of links.

Future Eve informs recent artworks, and in so doing continues to produce "artificial" offsprings. One such work is Javier Roca's *RE-constructing EVE*, an Extended Virtual Environment (EVE) commissioned for SIGGRAPH 99 Art Gallery (August 8–13, 1999). About his piece, Roca notes that this "topographic evocation of genetic engineering is ultimately a transitional work, an invitation to explore the 'multiplicity' and the complex relation between organism and machine, and hopefully, as in Villier's narrative text, reflects in this case a bridge between the twentieth century and the twenty-first century."[116] This piece, however, still works on a representational mode: it is, according to Roca, "a 'blue print,' an 'assemblage' of symbolic materials, interactions and historical anatomies of possible bodies." Less symbolic however, is the notion that there is already one new Eve, and that, as Villiers de l'Isle-Adam had prophesied, she was provided by science. Such is the premise of the Critical Art Ensemble's (CAE) *Cult of the New Eve* (CONE):

> When the Human Genome Project (HGP) began its mission of mapping and sequencing the entire human genome, it needed DNA in order to start . . . A review board with strict procedures was set up to insure the privacy of blood donors. However, after the first donor was approved, no other donors were needed. The DNA of the first approved volunteer was mass produced (copied) as needed. Why go to the trouble and expense of having any more? After all, one donor is sufficient for the project's needs. What is known about this donor is that she is a woman from Buffalo, New York. She is the Eve of the second genesis. It will be a curious sight to see if she, too, is labeled by science with the sign of origination.[117]

A rhetorical project that has given way to several performances in key nodes of the electronic art world (e.g., Zentrum für Kunst und Medientechnologie [ZKM] in Karlsruhe) or other venues (e.g., the streets of Brussels) since the year 2000, CONE is above all a discursive construction: in the classic vein of one of the most outspoken artistic collectives worried about the new wonders of biotechnologies, CONE is a parody of a religious capitalist ritual enacted for various audiences (a *profanation*, as Agamben would say). It translates critically most of Villier's insights and rephrases them in catchy aphorisms such as "We can make Eden. Paradise now!" or "The New Eve is our own. She is global." CONE remains, however, a rhetorical project and should be stressed only as a backbone to the CAE's other interventions; as such, it does not include a bioartistic practice, only a clever discursive production (and the CAE now knows, to its own demise, the difference between discursive production and bioartistic practice: in America under the Patriot Act, the second can lead you to jail. Quite a shame!).[118] Closer to my interest here, I shall now focus on two early bioartistic projects where I could find the *Future Eve* genealogy present, albeit in a distorted way.

The first of these projects, Joe Davis's *Microvenus* (1996), is arguably one of the first, if not the first, bioartistic piece. Carried out with the technical help of molecular geneticist Dana Boyd at Jon Beckwith's laboratory at Harvard Medical School and at Hatch Echol's laboratory at University of California, Berkeley, the piece consisted in the encoding of an icon in the DNA of a bacterium. The title of the piece came from the specific icon that Davis chose to encode in the DNA of the bacterium, an ancient Germanic rune shaped in the resemblance of the female genitalia. Davis contends that "the graphic 'Venus' icon drafted for the Microvenus project was inspired by some of the oldest messages Homo sapiens have left for themselves (i.e., ten- to fifty-thousand-year-old 'Venus figurines') and partly by episodes of censorship that are now historically associated with 'scientific' attempts to create messages for extraterrestrial intelligence."[119] About this piece, Adam Zaretsky has noted that "these sequences were chosen by Davis to exemplify a certain aesthetic and that his aesthetic is not expressed visibly by the organisms in question. Instead, the message is genomically embedded poetic license, without gene function and presumably without any organismic effect."[120]

Of Villier de l'Isle-Adam's original project remains the idea that the "copy"—here in the renewed meaning of the palimpsest of life—might outlive the human original, and that the new Eve (or one of his alter egos here, i.e., Venus) is the meaningful message of an obsolete humanity. Quite paradoxically, Davis's encoded icon is actually invisible and, even more important, without (direct) phenotypical effect on the bacterium that "carries' it: it is, in other words, *junk DNA*. In some ways, the icon is an invisible message, the invisible message of a New Genesis, and the link moves from the realm of the visible to the realm of the readable.

This idea of a new Genesis is also developed in the piece of another bioart pioneer—transgenic art, in his own words—in Eduardo Kac's piece aptly named *Genesis* (1999). Commissioned by Ars Electronica and presented online and at the O.K. Center for Contemporary Art, in Linz, Austria, from September 4 to September 19, 1999, this transgenic artwork consists of yet another inscription in bacterial DNA, what Kac calls an "artist's gene." Kac created this synthetic gene (in fact, he commissioned it to scientists who actually created it) by "translating a sentence from the biblical book of Genesis into Morse Code, and converting the Morse Code into DNA base pairs according to a conversion principle specially developed by the artist for this work."[121] The sentence reads: "Let man have dominion over the fish of the sea, and over the fowl of the air, and over every living thing that moves upon the earth" (Genesis 1:28). *Overman redux.*

Kac designed the piece so that "participants on the Web could turn on an ultraviolet light in the gallery, causing real, biological mutations in the bacteria. This changed the biblical sentence in the bacteria. The ability to change the sentence is a symbolic gesture: it means that we do not accept its meaning in the form we inherited it, and that new meanings emerge as we seek to change it."[122] He insisted that he had chosen Morse code because, "as the first example of the use of radiotelegraphy, it represents the dawn of the information age—the genesis of global communication."[123] He could have added that he was being true to one of the original insights at the origins of molecular biology, Erwin Schrödinger's intuition.

So, in this piece at least, it is obvious that the new or second Genesis is (also) the genesis of a new age, that is, of the new kind of capitalism that I have dubbed, after Deleuze and Guattari, *capitalism of the fourth kind.* Under this renewed reign of the *Nexus,* artists still have to demonstrate how they can escape the rigors of what the theoreticians of the Frankfurt School had called "integration" and that we can now more aptly call *recycling.* From the days of the latest short-lived revolution—the beautiful spring of 1968—capitalism has shown without mercy that it can, indeed, recycle its fiercest critics, to the point that *The Commentaries on the Society of the Spectacle* now appears to be the new bible of the communication VP and other advertising agencies executives.[124]

When Davis and Kac created "artist genes" and encoded them into the DNA of another form of life (be it a bacterium, a plant, or an animal), they apparently did very different things. Indeed, they used the same technique (recombinant DNA). They both created an intermediary, a vector that molecular biologists call a "plasmid": a circular double-stranded DNA molecule (separate from chromosomal DNA) capable of autonomous replication, a vector, or, in Davis's words, "a virus-like entity."[125] In both cases, these plasmids encoded a meaningful message for the human experimenter/artist: a sentence from the Bible or a German rune. But their works seem to vary tremendously according to the point of insertion of their vectors.

Davis chose to insert his vector to no phenotypical effect, and, in fact, introduced more junk into the host DNA. By an ironic twist, the "meaningful message" that he wanted to introduce actually amounts to more junk for the host. In other words, since this encoding does not alter the functioning of the coding DNA that it transforms, the introduction does not result in a different protein synthesis but rather piles up with the noncoding DNA of the host, that is, its *junk DNA*. Materially speaking, the host is not altered, and this is why Zaretsky speaks of an "invisible" aesthetics. *Readable, but not visible; (in)-significant junk.*

Kac, on the other hand, chose to insert his "artist gene" into the coding part of the DNA. In another of his pieces, titled *Move 36*, he coupled his artist gene (in this case the Cartesian cogito) with a functional gene, that is, a gene with a phenotypical effect:

> "Move 36" makes reference to the dramatic move made by the computer called Deep Blue against chess world champion Gary Kasparov in 1997 . . . The installation presents a chessboard made of earth (dark squares) and white sand (light squares) in the middle of the room. There are no chess pieces on the board. Positioned exactly where Deep Blue made its Move 36 is a plant whose genome incorporates a new gene that I created specifically for this work. The gene uses ASCII . . . to translate Descartes's statement: "Cogito ergo sum" (I think therefore I am) into the four bases of genetics. Through genetic modification, the leaves of the plants curl. In the wild these leaves would be flat. The "Cartesian gene" was coupled with a gene that causes this sculptural mutation in the plant, so that the public can see with the naked eye that the "Cartesian gene" is expressed precisely where the curls develop and twist.[126]

In his piece titled *Genesis*, he took yet another strategy: he enabled the online visitors to the installation to voluntarily mutate the transcoded bacteria with an interactive interface that activated an ultraviolet light. And Adam Zaretsky seems to aptly conclude:

> Instead of emphasizing a permanent, hereditary thumbprint, a sort of "artist was here" designer organism, *Genesis* emphasized the continued evolution of transgenic living organisms beyond the intentionality of the artist's hands. Though the emphasis on codex and genetic code have their similarities with previous transgenic works, Eduardo Kac inserts not a mythic signature of genetic graffiti alone, but a living text which is subject to environmental degradation, popular mangling, multiple re-readings and continued mutant alterity.[127]

Note, however, that the only way for Kac to produce (i.e. to master) a visible effect is (1) in the case of *Move 36*, by coupling his artist gene to a "ready-made gene," one that is known to be functional, and (2) in the case of *Genesis*, to use a mutagenic agent (i.e., UV light). Thus, in themselves, his "artist genes" do not differ essentially from those crafted by Davis. They too are meaningful junk. Again, on an interesting new twist on the history of art, the visible has given way to the readable, and in both cases, the aesthetic posture requires the explanatory discourse of the artists. The real cyborg, the artificial creature, is, in both cases, invisible to the piece's audience: it is the plasmid, this virus-like entity, that is the true creation.

Reduced to its smallest living/nonliving components, the virus, individuated code in swarms, a multitude of them, in symbiotic, parasitic, and/or genetic associations, life appears eventually as always-already-a-Nexus. And the question of the human becoming becomes itself the site of recombinant practices that might, of course, displace a few frontiers (visible/readable, for instance) but that also works according to the same logic that already was: the connective logic of life as a form of association. That, in the process, the symbolic turns to junk and junk to a site of potential redemption should not come as a surprise. That, in the process, the meaningful turns into the invisible (and vice versa), should not either. Remember, life from the start was recast as Nexus, both relation and exploitation, as in a twisted master-and-slave dialectic: Carbon and Silicon, Kaczynski and microVenus, Dekard and Rachel, Edison and Halaly. Who's the slave, who's the Master? *Who's in control anymore??*

But before we come back to this most troubling question, let us see yet another strategy employed to come to terms with it: instead of descending the phylogenic ladder (from Human to Eukaryotes and down to Prokaryotes, bacteria, and viruses), let us move backwards on the ontogenic staircase, from the body to its organs, from the Body without Organs (BwO) to the Organs without Body (OwB), its obverse. These two strategies somehow converge, or, better said, are two modalities of the same phenomenon, since it is one of the often alleged (but probably false) laws of biology that *ontogeny recapitulates phylogeny* (one climbs the ladder in order to descend the stairway, or vice versa). In this perspective, Slavoj Žižek is right to write that there is "a trend" in today's science and technology that both makes emerge a "body in pieces" *and* culminates in the biogenetics notion that "the true center of the living body is not his soul but its genetic algorithm."[128] Unsurprisingly, to shift from one modality to the other is also a passage that current bioartistic practices have taken.

I refer here specifically to the work of the *Tissue Culture and Art Project* (TC&A hereafter), alone or in association with Stelarc. Ionat Zurr and Oron Catts, the two artist members of TC&A, consider that their work, involving "the manipulation of living tissues outside and independent to the organism they were derived from," provides an alternative to the kind of manipulations practiced currently in molecular biology protocols. They insist that "artists dealing with genetics consider the genetic code in a similar way to the digital code. As a result the manipulation of life becomes 'manipulation of a code.'" That is why, if their art belongs to contemporary bioart as we defined it, it cannot be included in the less general category of "transgenic art" (as in the work of Davis and Kac discussed earlier). They add that the epistemological and ethical questions raised by their artistic interventions are not addressed by existing discourse, because "the manipulation of tissues is visceral."[129]

Indeed, their work involves the production (culture) of tissues or neo-organs, and they agree that it is "about producing body spare parts."[130] Trained in both arts and biological experimental protocols necessary to their practice, they do not ignore the philosophical references that made me consider these spare parts as necessary components for the machine of the fourth kind.[131] Accordingly, they wonder about the ethical and political consequences with these spare parts, in which they see an instance of what they call "the Semi-Living" or "Partial Life":

> Working with the Semi-Living and Partial Life, we are confronted with the question; are we creating another form of life for exploitation? . . . in the long term, they [the semi-living and partial life entities] confront the viewer with the realization that life is a continuum of the different metabolizing beings and in the transition from life to death, and from the living to the non-living. Their existence contradicts the conventional dichotomies that govern traditional and current Western ethical systems.[132]

I shall get back to the full extent of the consequences of this posture when I return to the golem. But, in the meantime, let us remember that it is quite a contemporary posture, that of *hylozoism:*

> Everything that exists, the whole of Nature is alive—it suffers and enjoys. There is no death in this universe; what happens in the case of "death" is just that particular coordination of living elements disintegrates, whereas Life goes on, both the Life of the Whole and the life of the elementary constituents of reality . . . We find this position from Aristotle (his notion of soul as the One-form of the body) . . . up to the whole panoply of today's theories, form the notion of Gaia (Earth as a living organism) to Deleuze, the last great philosopher of the One, the "body without organs" that thrives in the multitude of its modalities.[133]

"The bond that humans share with all living beings" is TC&A's version of the Nexus. It is both the axiom and the result of their strategy involving "the phylogenic staircase," my second characterization of contemporary bioartistic practices of the Human Nexus. But in their case, one more degree of complexity arises from the fact that the Nexus is distributed across an interface that is also part of it: *Nexus square, metanexus, if you will.* The nexus as process is indeed squared when it is both the matrix (the artificial womb) and the form of life that grows inside.

The embodiment of the human becoming by bioartistic discourses and experimentations instigates a new conception of the human body that participates in its dis-/re-embodiment. This notion is central to our understanding of bioart: along with discourses, bioartists engage in a global reflection that participates in human culture that might also be anchored into the biological dogma. Simultaneously, artists experiment with living systems, tissues or nucleic acids, and begin to challenge their given uses, purposes, and meanings. Bioartists thus create pieces that link computational systems (hardware and software) and organic matter (wetware), sometimes creating hybrids or chimeras, monstrous or invisible (albeit readable) effects, all belonging to what I call *junkware.* In so doing, they participate in the production of a body that is both, in fact, a new body and a reconfiguration of the original ("natural") body. Thus, articulating (discursively and experimentally) the body in its singular artificial organs is simultaneously participating in its disarticulation—or dislocation—with the natural body and initiating a reflection on its potential reconfigurations. The Tissue Culture and Art Project initiates a corporeal disruption between the inside and the outside, and reveals a higher power of the Nexus, on both sides of the interface that it artificially creates: cells and tissues, organ and matrix, world and human: one Nexus (talk about panpsychism!).

I have left waiting the figure of the servant, this second image of the android. It is now time for him to come back (with a vengeance). But behind the servant android lures the golem of legend, and that, we feel, is dead end, because, again, IT IS NOT ABOUT CONTROL: today's Nexus is beyond control. Or, more exactly, one has to realize that the servant golem is but one side of the golem, its emanation on only one plane of his two constitutive planes:

> The Golem has always existed on two quite separate planes. The one was the plane of ecstatic experience where the figure of clay, infused with all those radiations of the human mind, which are the combinations of the alphabet, became alive for the fleeting moment of ecstasy, but not beyond it. The other was the legendary plane where Jewish folk tradition, having heard of the Kabbalistic speculations on the spiritual plane, translated them into down-to-earth tales and traditions . . . The Golem, instead of being a spiritual experience of man, became a technical servant of man's needs, controlled by him in an uneasy and precarious equilibrium.[134]

It is to this second plane of the ecstatic experience that I would like to draw your attention now. Sonya Rapoport, in her redemption of Eduardo Kac's *Genesis* gene, has caught a glimpse of this plane. Her Web work titled "Redeeming the Gene, Molding the Golem, Folding the Protein" is a mythic parody that challenges Kac's work with the creation of a golem, brought to life according to the Jewish esoteric practices of Kabbalah.[135] According to Rapoport, Kac's artist gene needs redemption because of the way it was produced, or, more accurately, because of the languages (codes) of his making: Kac is guilty of having used the King James translation of the Bible (rather than the Hebrew text of the Torah) and Morse's code (and Morse was pro-slavery). At the opposite, her golem is a positive force, brought to life by two women (Eve, of course, and her Gnostic alter ego, Lilith, "who irritated the Lord of Creation by demanding equal rights" [Scholem]). In the end, Kac is redeemed, and in one of the last screens, his face is replaced with that of Adam Kadmon, the Primordial Man of Lurianic Kabbalah. For Luria, Adam was twice a golem: he was first a gigantic golem *(Adam Kadmon)* and second an ordinary golem *(Adam Rishon):*

> Man, as he was before his fall, is conceived as a cosmic being which contains the whole world in itself and whose station is superior even to that of Metatron, the first of the angels. *Adam Ha-Rishon,* the Adam of the Bible, corresponds on the anthropological plane to *Adam Kadmon,* the ontological primary man. Evidently the human and the mystical man are closely related to each other; their structure is the same, and to use Vital's own word, the one is the clothing and the veil of the other. Here we have also the explanation of the connection between man's fall and the cosmic process, between morality and physics. Since Adam was truly, and not metaphorically, all-embracing, his fall was bound likewise to drag down and affect everything, not merely metaphorically but really. The drama of *Adam Kadmon* on the theosophical plan is repeated, and paralleled by that of *Adam Rishon.*[136]

By her use of the Lurianic Kabbalah (for some other choices were indeed possible), Rapoport reinforces the Gnostic emphasis of her piece, but she also gives us a very contemporary key to unlock the Nexus.

Today's Human Nexus, and his associated Second Genesis, is the Eternal Return of the Primordial Man. As in the first time around, the question raised is that of his freedom.

De-Coda

> And we: onlookers always, over all,
> Interested in everything and never looking out!
> Overfills us. We order. It decays.
> We order again and ourselves decay.
> Who turned us around so we,
> No matter what have the pose
> Of one who is departing? As he who on
> The last hill, which still shows
> His whole valley will turn, halt, pause—
> So we live, forever taking leave.
>
> —RAINER MARIA RILKE, "The Eighth Elegy"

The charitable sweepstake of creating this culture of junk is sponsored by a random computing ballot system, as matter without even a cause anymore. Junk, this culture of the perpetual potentiality of consumption, is the nostalgic hope of the prize recipient in stacks, piles, drawers and governmental parastatals. It materializes the last humans before the minimum one tenth of creating this incarnation of consumption that might regenerate our worst food. It is thus no meaningless coincidence that all participants were selected from a glimpse of the allegedly useless part our genetic inheritance. You forget about it, and of your locality, after the nostalgic hope without even a group of creating this culture of matter of principle, you are to donate at the falling dust for it somehow grows anarchically. It might be that these processes are quasi-necessary; they deliver a cashier's fate. It is our pleasure to donate at the nostalgic hope of corporate organizations and cellars. This invariably means that what might regenerate our whole culture has indeed become virtual. House, choose memories from principle, before system. On the marginal living spaces of the above ballot system, junk comes in the making, and rusts, fades, decays. It is my pleasure to process your e-mail address attached to come.

Congratulations once more.

Faithfully yours,

Mrs. Sophie Dick (sophie02@saintly.com)

Sib **Tripping over the Organism**

Richard Doyle once reported that "there were [in the 1950s and 1960s] many experiments among researchers attempting to ingest DNA and RNA itself as a hallucinogen, sometimes in the hope of developing a 'learning lozenge' which would inscribe the experience of LSD onto the brain."[1] At the same time, these "researchers" were, quite often, albeit unwillingly, working for the CIA. Quite a (power) trip! This is folklore by now, raw material for constipation theorists. No, Doyle had some more important revelations to make. These researchers, such as Kary Mullins, the inventor of polymerase chain reaction (PCR), were pretty much in tune with both Narby's shamans and Richard Dawkins's selfish programs. And the psychedelic gurus themselves shared this esoteric knowledge:

> These recipes and techniques for ecstasy were repeatedly and explicitly linked by Leary to the writing and execution of a sequence of steps in a computer environment, programming . . . The aim would be to lead the voyager to one of the visions deliberately, or through a sequence of visions . . . One can envision a high art of programming psychedelic sessions, in which symbolic manipulations and presentations would lead the voyager through ecstatic visionary Bead Games. In this framework, hallucinogenic subjects become both authors of and platforms for "symbolic manipulations and presentations," interactive wetware of infinite experiment and transformation. In the preface to a later work, Leary would write that "The Psychedelic Experience was our first attempt at session programming."[2]

This rhetoric is a two-way equivalence: DNA is a spirit is a drug is a program. DNA is both medium and message: one only has to "turn on, tune in, and drop out," as Leary vigorously engaged every young hippie (following McLuhan's advice to come up with "something snappy" to promote LSD). Turn on the computer, tune in the multimedium, that is. Not everybody dropped out, though. Some might have even done better than that: conquer power and authority by means of turning on and tuning in. Hear this rumor that has traveled the Web endlessly, starting only a week after the death of the Pope of Molecular Biology himself: "Francis Crick, the Nobel Prize-winning father of modern genetics, was under the influence of LSD when he first deduced the double-helix structure of DNA."[3]

So, a psychedelics-induced vision could somehow have become the key of success for a whole generation of aspiring "revolutionaries" (some found yesterday on Apple's ads of the "think different" series). The computer itself soon became both means and ends of this countercultural revolution of sorts, and its history was promptly rewritten as such, garage stories and gentle politically correct fables before the term even existed (a true revolution, then). Quite naturally, the personal development movement of the same period thrived on such a massive "reprogramming": how to make a business out of a well-timed metaphor. Timing is everything, and for the rest, there is LSD. No, scratch that: and for the rest there is some better software. Everybody became an accomplice in a gigantic cover-up: business as usual, capitalism recycling its "critics." Or, as Debord noticed, the society of the spectacle "defines the program of a ruling class and presides over its constitution. It presents the pseudo-goods to desire as well as it offers to the local revolutionaries the false models of revolution."[4] Or, with even more precision: "Everyone today wallows in the molecular as they do in the revolutionary . . . we should not, however, rediscover as an apparatus *(dispositif)* of desire what the cyberneticists have described as a matrix of code and of control."[5]

To the rituals of humanity's childhood, we, poor moderns, have substituted a consumption ostentatiously forbidden for the wealth of all mafias; mafia of the righteous discourse in robe, representative mafia of the conservative discourse, mafia of the liberated who nevertheless don't inhale, mafia of the exploitation of all the junk dreams, et cetera. Be a revolutionary: just say no. Not a problem, we still have our DNA, and everybody's a shaman, right?

The doors closed, shrieking; alas, from now on we are all mules, my dear Aldous.

xis **Thinking Junk and Period Pieces**

> Man is but a reed, the most feeble thing in nature, but he is a thinking reed. The entire universe need not arm itself to crush him. A vapor, a drop of water suffices to kill him. But, if the universe were to crush him, man would still be more noble than that which killed him, because he knows that he dies and the advantage which the universe has over him, the universe knows nothing of this. All our dignity then, consists in thought. By it we must elevate ourselves, and not by space and time which we cannot fill. Let us endeavor then, to think well; this is the principle of morality.[6]

Pascal, this wise, albeit uptight, man, knew it already. Man is but thinking junk, pondering rush. Today, everybody seems to agree that rush is both our essence and our *medium*, and ecology rules.

> Rubin, in some way that no one quite understands, is a master, a teacher, what the Japanese call a *sensei*. What he's a master of, really, is garbage, kipple, refuse, the sea of cast-off goods our century floats on. *Gomi no sensei*. Master of junk . . .

> He has nothing to say about *gomi.* It's his medium, the air he breathes, something he's swum in all his life.[7]

Our junk culture is Philip K. Dick's legacy; we are junkware, and the question of theodicy is first. This junk world begs Gnosticism, which is so often obvious these days. Remember: this world is disjunked, its understanding might require a non-Aristotelian logic such as the one imagined by Alfred Korzybski (in this sense, K. conceptualized our world,[8] and PKD wrote it into being). K., a man of his time not devoid of qualities, is the too-often-occulted influence of at least one of the original version of cybernetics, that of the late Gregory Bateson *(the map is not the territory, the word is not the thing, and yet, we are our own epistemology).* Bateson founded a cybernetic anthropology. K., a man of no time devoid of qualities, is the often-occulted influence of at least one of the original versions of contemporary science fiction, that of the cyberpunks. Cyberpunks enacted a cybernetic anthropology. But cyberpunk is an afterthought, a mere lexical revamping and technological fix to a creation that spanned the two previous decades: incipit *junkware.* Cyberpunks are the illegitimate (i.e., punk) children of Dick and Burroughs, Derrida and Kaczynski.[9]

Junk materializes events and binds them together in a rhizomatic fashion, in Markov chains. The event is the basic unit of cybernetics, hence the cinematographic presence of junk in cyberfictions. Thus van Vogt's *Voyage of the Space Beagle* and PKD's *Do Androids Dream of Electric Sheep?* gave rise to *Alien* (1979) and *Blade Runner* (1982), respectively, two of the most successful science-fiction movies of the late twentieth century, both directed by Ridley Scott. In both cases, the focus is on an evolutionary process reaching its limit, facing the Android and the Alien. In both, the human species is junk, oscillating between decay and bare survival. Junk is the presence of the terminal identity, exploded into myriad fragments, doubting its own presence. Clinging to life. Junk is the artificial memory of the replicant, the desperate survival of the living.

In cyberfictions, the fabric of time is disjunct, or, to put it in slightly different words, these fictions operate on a disjunctive temporality. This disjunctive temporality is the cybernetic time, the statistical time of "time series," "discrete or continuous sequences of measurable events distributed in time." In 1948, Norbert Wiener wrote: "Time series can in no way be reduced to an assembly of determinate threads of development in time," but rather must be seen as evolutionary processes. Around 1945, this world entered in the absolute yet uncertain time of the parallel universes, forever illuminated by the gravity's rainbow of the atomic bombs, floating over the ashes of the extermination camps. Homo sapiens then broke a threshold in its dual evolutionary process, hominization: he learned to drop the bomb and started his own genocide—recycling positive and negative feedbacks in singular ways. Both events disrupted the very fabric of time as he knew it. The first started a race that could provide the means for the total annihilation of his species. The second perfected eugenics to a level where it could *program* the total annihilation of his species, its autogeddon. Between these events and our awareness of their consequences lags the gestation of Homo nexus, the concluding term, so far, of the hardware-software-wetware series. Junkware, indeed, is also a period piece.

attP **Molecular Gods**

The molecular has become a religious affair, from the biologists to the philosophers who embraced it. In fact, it became the one religious "dogma" held by various people who often have atheism as their only common credo (for atheism is indeed a credo, albeit a negative one). Here is what Francis Crick said when Horace Freeland Judson asked him how and why he coined the expression "the central dogma":

> It was because, I think, of my curious religious upbringing . . . Because *Jacques* [Monod] has since told me that a dogma is something which a true believer *cannot doubt*! . . . But that *wasn't* what was in *my* mind. My mind was, that a dogma was an idea for which there was no *reasonable evidence*. You see?! . . . I just didn't know what dogma *meant*. And I could just as well have called it the "Central Hypothesis," or—you know. Which is what I meant to say. Dogma was just a catch phrase . . . It *is* sort of a super-hypothesis. And it's a *negative* hypothesis, so it's very very difficult to prove. It says certain transfers *can't* take place . . . The central dogma is much more powerful [than the sequence hypothesis], and therefore in principle you might have to say it could never be proved. But its utility—there was no doubt about that. Because if you *didn't* believe that, you could invent theories, *unlimited* theories, whereas if you just put in that one assumption, that once the sequence information had got into the protein it couldn't get out again, well then, essentially you were on the right track you see.[10]

I have shown how this hypothesis, if it was very productive for a while, turned later into a real "dogma," and thus prevented (for another while at least) any further inquiry along lines of thought deemed "counterproductive." The molecular dogma thus became what its name meant from the start. This is far from being the only instance of the religious becoming of the molecular, though. In fact, religious issues, debates, and controversies have regularly traversed the history of recent biology (and especially since World War II, and its cybernetics translation in the guise of molecular biology). The most crucial and also the latest instance of such a debate is, of course, the evolution–creation debate, or, in today's terminology, the Darwinist–intelligent design (ID) argument. Such a debate is in fact aporetic, since cybernetics and hence molecular biology referred to a renewed version of the teleological argument (under the "new guise" of *teleonomy*).

Moreover, the question of design is creeping through the back door in yet another paradoxical way. At the same time that most biologists, true to their rationalist and mechanist inheritance, strongly oppose any mention of a potential transcendental designer (or watchmaker or great architect or whatever name you please to grant him), two other major cultural phenomena affect the issue: (1) the overall Gnostic turn that culture, and especially cyberculture, has taken in the last thirty years or so ("the world PKD made"), and (2) the phantasm, the desire, or maybe even the opportunity for mankind to actually design its "successor" (see Homo nexus), or, more prosaically, for parents to design their babies.[11] In other words: no design if God could be the subject, but all design if we can take his place (over his dead body).[12] The problem is not design, but the designer: the Gnostic script has been reworked to accommodate us in the part of the demiurge (thanks to the molecular spin doctors).

Yet, in other words: it is a question of agency, and maybe even more a question of *will*. There are no surprises, then, in the return of the Promethean myth to give some sense to the whole issue. But, by a peculiar inversion, Prometheus, who eventually got tied up for his transgression, could also be seen as the tying god, and hence as the junk god. For junk is what ties us up to an ever-potential becoming, but also what ties us up to an uncertain future: Prometheus, true to his name, is a technological god all right, the god of anticipation. The engineer is, by vocation, somebody who will seek to master an uncertain future. Any technology is, to a certain extent at least, a system of anticipation, and materializes a teleological worldview. But among the technologies, some are squarely so: the mantic (or prophetic) technologies.

When the religious is disqualified, the mantic remains. The will to self-design supposes some sense of anticipation: the science of heredity turned upside down. The cliché that states that whoever forgets history is bound to repeat it has found a new use: as a motto for personal development, at the level of the whole species. Molecular biology is also a kind of mantic technology: given this instance of a "genetic message," given a normal functioning of the "genetic program," one should obtain a certain result (that given body, and maybe even that certain mind). Or, in other words, *a given genetic capital should yield a given personal revenue (in living money)*. Genetic determinism is today's credo for brand-new religions, new kinds of cargo cults derived from the central dogma: transhumanists, extropians, and so on. It aspires to become the state religion of capitalism of the fourth kind. Man™ was but an open-source problem with a design solution . . . er, sorry, Overman™ is all but an open-source solution to a design problem. Who ends up writing the copyright notice, or else the open-source copyleft license, for the Successor of Man™? is the question of the day.

Alas, then came junk, and this seducing picture of a well-mastered future now seems out of focus. Necessity, yes, but chance too. Chance? Are you sure?

int **Vanishing Sequences (End Credits)**

> The generation of numerous intercalated transcripts spanning the majority of the genome has been met with mixed opinions about the biological importance of these transcripts. Our analyses of numerous orthogonal data sets firmly establish the presence of these transcripts, and thus the simple view of the genome as having a defined set of isolated loci transcribed independently does not seem accurate. Perhaps the genome encodes a network of transcripts, many of which are linked to protein coding transcripts.[13]

So much for the famous scientific modesty. It means: central dogma, phenotypic paradigm, *Rest in Peace.* I was writing the last words of my manuscript when I heard the news of this "first-class funeral." I had the definite feeling that part of my subject matter was disappearing. There is no more junk on DNA, in this sense, if you still believed that junk is some sort of trash (vestige, fossils, etc.). Or, alternatively, DNA is (mostly) junk, if you understand that junk is the always present potentiality of a renewed function.

If "the simple view of the genome as having a defined set of isolated loci transcribed independently" (or as being "a sea of reverse-transcribed DNA with a small admixture of genes" [see chapter 1]) does not hold anymore, it practically means that the molecular biology notion of the gene is obsolete.[14] A fortiori, the simple semiotic view of a DNA sequence as an isolated sign makes little sense either (if you pardon me the pun). To isolate DNA sequences, or mRNA sequences, as isolated signs is not only no longer required, but would actually become "counterproductive."[15]

Michel Houellebecq was only partly right (or partly wrong, if you prefer): *the next mutation will be genetic because it is metaphysical.* Down with this "pharisaic demiurge," "who contemplates all uncreated possible worlds to take delight in his own single choice"![16] Unlike him, we ought to open our ears "to the incessant lamentation that, throughout the infinite chambers of this Baroque inferno of potentiality, arise from everything that could have been but was not, from everything that could have been otherwise but had to be sacrificed for the present world to be as it is."[17] Let us attune our ears again to the clamors of being. No more single choices, no more present world as it is, but many compossible and incompossible worlds patiently awaiting their anamnesis, not only "the memory of the time in which man was not yet man,"[18] but also this memory of a time to come in which overman was not man anymore.

Glossary

The sources of this glossary are, unless stated otherwise, Mary Chitty, *Genomics Glossaries and Taxonomies*, Cambridge Healthtech Institute, glossary last updated August 6, 2004, homepage last revised June 1, 2004, http://www.genomicglossaries.com/default.asp; *Introductory Biology Courseware- Glossary*, http://tidepool.st.usm.edu/crswr/glossaryen.htm; and *Genetics Glossary*, http://helios.bto.ed.ac.uk/bto/glossary.

adaptation In the evolutionary sense, some heritable feature of an individual's phenotype that improves its chances of survival and reproduction in the existing environment.

Alu family A dispersed intermediately repetitive DNA sequence found in the human genome in about three hundred thousand copies. The sequence is about three hundred bp long. The name Alu comes from the restriction endonuclease AluI that cleaves it.

amino acid The basic building block of proteins (or polypeptides). Containing a basic amino (NH2) group, an acidic carboxyl (COOH) group, and a side chain (R—of a number of different kinds) attached to an alpha carbon atom.

bacteriophage (see phage λ) A virus that infects bacteria.

base (or nitrogen base) Type of molecule that forms an important part of nucleic acid, composed of a nitrogen-containing ring structure. Hydrogen bonds between bases in opposing complementary strands link the two strands of a DNA double helix.

base pair (bp) Two bases that form a "rung of the DNA ladder." In base pairing, adenine always pairs with thymine, and guanine always pairs with cytosine.

purines A type of nitrogen base; the purine bases in DNA and RNA are adenine (A) and guanine (G).

pyrimidine Nitrogenous bases of which thymine (T) is found in DNA; uridine (U) in RNA; and cytosine (C) in both.

chromatid One of the two side-by-side replicas produced by chromosome replication in mitosis or meiosis. Subunit of a chromosome after replication and prior to anaphase of meiosis II or mitosis. At anaphase of meiosis II or mitosis when the centromeres divide and the sister chromatids separate, each chromatid becomes a chromosome.

chromatin The mixture of DNA and proteins, especially histones, that comprises eukaryotic nuclear chromosomes.

euchromatin The more open, unraveled form of eukaryotic chromatin, which is available for transcription. A chromosome region that stains poorly or not at all; thought to contain the normally functioning genes. Region of eukaryotic chromosome that is diffuse during interphase. Presumably the actively transcribing DNA of the chromosomes.

heterochromatin Densely staining condensed chromosomal regions, believed to be for the most part genetically inert. Nontranscribed eukaryotic chromatin that is so highly compacted that it is visible with a light microscope during interphase.

chromosome A linear end-to-end arrangement of genes and other DNA, sometimes with associated protein and RNA. The form of the genetic material in viruses and cells. A circle of DNA in prokaryotes; a DNA or an RNA molecule in viruses; a linear nucleoprotein complex in eukaryotes.

codon A section of DNA (three nucleotide pairs in length) or RNA (three nucleotides in length) that codes for a single amino acid. A sequence of three RNA or DNA nucleotides that specifies (codes for) either an amino acid or the termination of translation.

anticodon The three-base sequence in tRNA complementary to a codon on mRNA. A nucleotide triplet in a tRNA molecule that aligns with a particular codon in mRNA under the influence of the ribosome, so that the amino acid carried by the tRNA is added to a growing protein chain.

diploid The state of having each chromosome in two copies per nucleus or cell. A cell having two chromosome sets, or an individual having two chromosome sets in each of its cells.

discrete-time signal A time series, perhaps a signal that has been sampled from a continuous-time signal. Unlike a continuous-time signal, a discrete-time signal is not a function of a continuous-time argument, but is a sequence of quantities; that is, a function over a domain of discrete integers. Each value in the sequence is called a sample.

DNA (deoxyribonucleic acid) An antiparallel double helix of nucleotides (having deoxyribose as their sugars) linked by phosphodiester (sugar-phosphate) bonds to adjacent nucleotides in the same chain and by hydrogen bonds to complementary nucleotides in the opposite chain. The fundamental substance of which genes are composed.

complementary DNA (cDNA) This is DNA synthesized from a mature mRNA template by the enzyme reverse transcriptase. cDNA is frequently used as an early part of gene cloning procedures, since it is more robust and less subject to degradation than the mRNA itself.

jumping DNA See *transposon*.

noncoding DNA (ncDNA) Introns, spliced out of the messenger RNA following transcription. Noncoding DNA (also known as selfish, ignorant, parasitic, and incidental DNA) includes introns, transposable elements, pseudogenes, repeat elements, satellites, UTRs, hnRNAs, LINEs, SINEs, as well as unidentified junk and makes up approximately 97 percent of the human genome. Some scientists were so overwhelmed by the amount of noncoding DNA that they referred to the genome as "a collection of non-coding regions interrupted by small coding regions."[1]

recombinant DNA A novel DNA sequence formed by the joining, usually in vitro, of two non-homologous DNA molecules.

satellite DNA DNA that forms a separate band in a buoyant density gradient because of its different nucleotide composition (A:T rich DNAs are less dense than G:C rich DNAs). Highly repetitive eukaryotic DNA primarily located around centromeres. Satellite DNA usually has a different buoyant density than the rest of the cell's DNA.

selfish DNA A segment of the genome with no apparent function other than to ensure its own replication.

embryonic stem cells (ES) Cultured cells derived from the pluripotent inner cell mass of blastocyst-stage embryos.

enhancer A short segment of genomic DNA that may be located remotely and that, on binding particular proteins (transacting factors), increases the rate of transcription of a specific gene or gene cluster.

eukaryote A life-form comprised of one or more cells containing a nucleus and membrane-bound organelles. Included are members of the Kingdoms Protista, Fungi, Plantae, and Animalia.

exon The parts of a genetic transcript remaining after the introns are removed and which are spliced together to become a messenger or structural RNA. A region of a gene that is present in the final functional transcript (mRNA) from that gene. Any non-intron section of the coding sequence of a gene; together the exons constitute the mRNA and are translated into protein.

gamete A germ cell having a haploid chromosome complement. Gametes from parents of opposite sexes fuse to form zygotes. A specialized haploid cell that fuses with a gamete from the opposite sex or mating type to form a diploid zygote; in mammals, an egg or a sperm.

gene (cistron) Structurally, a basic unit of hereditary material; an ordered sequence of nucleotide bases that encodes one polypeptide chain (via mRNA). The gene includes, however, regions preceding and following the coding region (leader and trailer) as well as (in eukaryotes) intervening sequences (introns) between individual coding segments (exons). Functionally, the gene is defined by the cis-trans test that determines whether independent mutations of the same phenotype occur within a single gene or in several genes involved in the same function.

jumping gene See *transposon*.

pseudogene An inactive gene derived from an ancestral active gene. Sequences that are generally untranscribed and untranslated and which have high homology to identified genes. However, it has recently been shown that in different organisms or tissues functional activation may occur. Therefore, the previous policy of assigning the gene symbol of the structural gene followed by "P" and a number will only be approved on a case-by-case basis.[2]

These genes bear a close resemblance to known genes at different loci, but are rendered nonfunctional by additions or deletions in structure that prevent normal transcription or translation. When lacking introns and containing a poly-A segment near the downstream end (as a result of reverse copying from processed nuclear RNA into double-stranded DNA), they are called processed genes.

regulator gene A gene primarily involved in controlling another (structural) gene. Genes that are involved in turning on or off the transcription of structural genes.

structural gene A gene encoding the amino acid sequence of a protein. Nonregulatory gene.

gene interaction network A network of functional interactions between genes. Functional interactions can be inferred from many different data types, including protein–protein interactions, genetic interactions, co-expression relationships, the co-inheritance of genes across genomes, and the arrangement of genes in bacterial genomes. The interactions can be represented using network diagrams, with lines connecting the interacting elements, and can be modeled using differential equations.

gene silencing The switching off of a gene by an epigenetic mechanism at the transcriptional or posttranscriptional levels. Includes the mechanism of RNAi.

genetic code The sequence of nucleotides, coded in triplets (codons) along the mRNA, that determines the sequence of amino acids in protein synthesis. The DNA sequence of a gene can be used to predict the mRNA sequence, and the genetic code can in turn be used to predict the amino acid sequence.

> The notion of a "code" as the key to information transfer was not articulated publicly until late 1954, when [George] Gamow, Martynas Ycas, and Alexander Rich published an article that defined the code idiom for the first time since Watson and Crick casually mentioned it in a 1953 article. Yet the concept of coding applied to genetic specificity was somewhat misleading, as translation between the 4 nucleic acid bases and the 20 amino acids would obey the rules of a cipher instead of a code. As Crick acknowledged years later, in linguistic analysis, ciphers generally operate on units of regular length (as in the triplet DNA scheme), whereas codes operate on units of variable length (e.g., words,

phrases). But the code metaphor worked well, even though it was literally inaccurate, and in Crick's words, "'Genetic code' sounds a lot more intriguing than 'genetic cipher.'" Codes and the information transfer metaphor were extraordinarily powerful, and heredity was often described as a biological form of electronic communication.[3]

degenerate code A code in which several code words have the same meaning. The genetic code is degenerate because there are many instances in which different codons specify the same amino acid. A genetic code in which some amino acids may each be encoded by more than one codon.

genome The entire complement of genetic material in a chromosome set. The entire genetic complement of a prokaryote, virus, mitochondrion or chloroplast, or the haploid nuclear genetic complement of a eukaryotic species.

genotype The specific allelic composition of a cell, either of the entire cell or, more commonly, for a certain gene or a set of genes. The genes that an organism possesses.

haploid The state of having one copy of each chromosome per nucleus or cell. A cell having one chromosome set, or an organism composed of such cells.

histones The chief protein components of chromatin. They act as spools around which DNA winds and they play a role in gene regulation.

hybridization The process of joining (annealing) two complementary single-stranded DNAs into a single double-stranded molecule. In microarray analysis, the target RNA/DNA from the subject under investigation is denatured and hybridized to probes that are immobilized on a solid phase (i.e., glass microscope slide).

indel Insertion and deletion of DNA, referring to two types of genetic mutation. To be distinguished from a "point mutation," which refers to the substitution of a single base.

insertion Several nucleotides can be added to a sequence, resulting in an insertion. Effects of an insertion are variable. Insertions and deletions can be hard to tell apart and are sometimes referred to collectively as "indels."

intergenic DNA Any of the DNA in between gene-coding DNA, including untranslated regions, 5' and 3' flanking regions, INTRONS, nonfunctional pseudogenes, and nonfunctional repetitive sequences. This DNA may or may not encode regulatory functions.

interphase The cell cycle stage between nuclear divisions, when chromosomes are extended and functionally active. The metabolically active nondividing stage of the cell cycle.

intron An intervening section of DNA that occurs almost exclusively within a eukaryotic gene, but which is not translated to amino acid sequences in the gene product. The introns are removed from the pre-mature mRNA through a process called splicing, which leaves the exons untouched, to form an active mRNA.

Sequences of DNA in the genes that are located between the EXONS. They are transcribed along with the exons but are removed from the primary gene transcript by RNA SPLICING to leave mature RNA. Some introns code for separate genes.

A segment of DNA that is transcribed, but removed from within the transcript by splicing together the sequences (exons) on either side of it.

karyotype A photomicrograph of an individual's chromosomes arranged in a standard format showing the number, size, and shape of each chromosome type; used in low-resolution physical mapping to correlate gross chromosomal abnormalities with the characteristics of specific diseases.

Markov chain A discrete-time stochastic process with the Markov property, which means that every future state is conditionally independent of every prior state. At each time the system may have changed from the state it was in the moment before, or it may have stayed in the same state. The changes of state are called transitions.

meiosis Two successive nuclear divisions (with corresponding cell divisions) that produce haploid gametes (in animals) or haploid sexual spores (in plants and fungi) having one-half of the genetic material of the original cell. The nuclear and cell division process in diploid eukaryotes that results in four haploid gametes or spores having one member of each original pair of homologous chromosomes only per nucleus.

methylation A term used in the chemical sciences to denote the attachment or substitution of a methyl group (CH_3) on various substrates. This term is commonly used in chemistry, biochemistry, and the biological sciences. In biochemistry, methylation more specifically refers to the replacement of a hydrogen atom with the methyl group. In biological systems, methylation is catalyzed by enzymes; such methylation can be involved in modification of heavy metals, regulation of gene expression, regulation of protein function, and RNA metabolism. DNA methylation and proteins methylation are two crucial epigenetic phenomena.

microarray An arrayed set of probes for detecting molecularly specific analytes or targets. Typically, the probes are composed of DNA segments that are immobilized onto the solid surface, each of which can hybridize with a specific DNA present in the target preparation. DNA microarrays are used for profiling of gene transcripts.

microsatellites They consist of tandem repeats, which contain repetitive runs of the same short base sequence (e.g., GTA, GTA, GTA . . .). Among individuals, these sections of DNA may vary in the number of repeats they contain and can serve as markers and signs of genetic variation.

microtubules Hollow cylinders made of the protein tubulin that form, among other things, the spindle fibers.

mitosis The nuclear division producing two daughter nuclei identical to the original nucleus. A type of nuclear division that produces two daughter nuclei identical to the parent nucleus normally just prior to cell division.

mutation (1) The process producing a gene or a chromosome differing from the wild-type. (2) The gene or chromosome that results from such a process.

frameshift mutation A type of mutation in which a number of nucleotides not divisible by three is deleted from or inserted into a coding sequence, thereby causing an alteration in the reading frame of the entire sequence downstream of the mutation. These mutations may be induced by certain types of mutagens or may occur spontaneously.

point mutation A mutation that can be mapped to one specific site within a locus. A small mutation that consists of the replacement (transition or transversion), addition, or deletion (frameshift) of one or a few bases.

silent mutation A mutation in which the function of the protein product of the gene is unaltered.

spontaneous mutation A mutation occurring in the absence of mutagens, usually due to errors in the normal functioning of cellular enzymes.

nucleotide Subunit that polymerizes into nucleic acids (DNA or RNA). Each nucleotide consists of a nitrogenous base, a sugar, and one to three phosphate groups.

operon A functional unit consisting of a promoter, an operator, and a number of structural genes, found mainly in prokaryotes. The structural genes commonly code for several functionally related enzymes, and although they are transcribed as one (polycistronic) mRNA, each is independently translated. In the typical operon, the operator region acts as a controlling element in switching on or off the synthesis of mRNA (operator gene).

The genetic unit consisting of a feedback system under the control of an operator gene, in which a structural gene transcribes its message in the form of mRNA upon blockade of a repressor produced by a regulator gene. Included here is the attenuator site of bacterial operons where transcription termination is regulated.

parthenogenesis The production of offspring by a female with no genetic contribution from a male. The development of an individual from an unfertilized egg that did not arise by meiotic chromosome reduction.

phage λ Lambda phage is a virus particle consisting of a head, containing double-stranded linear DNA, and a tail. It was discovered by Esther Lederberg in 1950 and has been used heavily as a model organism in molecular biology. λ phage belongs to the "temperate" phages family—a term coined by Elie Wollman in 1952 in reference to J. S. Bach's famous musical composition—because its life cycle has two alternate pathways, the lytic pathway and the lysogenic pathway.

The phage injects its DNA into its bacterial host through the tail, and then usually enters the lytic pathway, where it replicates its DNA, degrades the host DNA, and hijacks the cell's replication, transcription, and translation mechanisms to produce as many phage particles as cell resources allow. When cell resources are depleted, the phage will lyse (break open) the host cell, releasing the new phage particles. However, under certain conditions, the phage DNA may integrate itself into the host cell chromosome in the lysogenic pathway. In this state, the λ DNA is called a prophage and stays resident within the host's genome without apparent harm to the host, which can be termed a lysogen when a prophage is present. The prophage is duplicated with every subsequent cell division of the host. The phage genes expressed in this dormant state code for proteins that repress expression of other phage genes. These proteins are broken down when the host cell is under stress, resulting in the expression of the repressed phage genes.

phenotype (1) The form taken by some character (or group of characters) in a specific individual. (2) The detectable outward manifestations of a specific genotype. (3) The observable attributes of an organism.

plasmid Autonomously replicating extrachromosomal DNA molecule. An autonomous self-replicating genetic particle usually of circular double-stranded DNA.

polymerase chain reaction (PCR) A molecular biology technique for replicating DNA in vitro. The DNA is thus amplified, sometimes from very small amounts. PCR can be adapted to perform a wide variety of genetic manipulations.

polypeptide A chain of linked amino acids; a protein.

polyploid A cell or an organism having three or more chromosome sets.

prokaryote A cell in which the genetic material is not contained within a nucleus (bacteria).

promoter Region on a DNA molecule involved in RNA polymerase binding to initiate transcription. A regulatory region a short distance upstream from the 5' end of a transcription start site that acts as the binding site for RNA polymerase. A region of DNA to which RNA polymerase binds in order to initiate transcription.

recombination (1) The nonparental arrangement of alleles in progeny that can result from either independent assortment or crossing over. (2) In general, any process in a diploid or partially diploid cell that generates new gene or chromosomal combinations not found in that cell or in its progenitors. (3) At meiosis, the process that generates a haploid product of meiosis whose genotype is different from either of the two haploid genotypes that constituted the original meiotic diploid.

repetitive sequences DNA made up of copies of the same or nearly the same nucleotide sequence. DNA sequences that are present in many copies per chromosome set. Repetitive DNA sequences may be closely linked (e.g., satellite DNA or VNTR loci) or dispersed throughout the genome or parts of the genome (e.g., Alu family). They make up at least 50 percent of the genome. Repetitive sequences are thought to have no direct functions, but they shed light on chromosome structure and dynamics. They hold important clues about evolutionary events, help chart mutation rates, and, by seeding DNA rearrangements, can modify genes and create new ones. They also serve as tools for genetic studies.

The vast majority of repeated sequences in the human genome are derived from transposable elements—sequences like those that form viral genomes—that propagate by inserting fresh copies of themselves in random places in the genome. A full 45 percent of the human genome derives from such transposons. A major surprise of this new global analysis of the human genome is that many components in this diverse array of repeated sequences, traditionally considered to be "junk," appear to have played a beneficial role over the course of human evolution.[4]

interspersed repetitive sequences Copies of transposable elements interspersed throughout the genome, some of which are still active and often referred to as "jumping genes." There are two classes of interspersed repetitive elements. Class I elements (or RETROELEMENTS—such as retrotransposons, retroviruses, LONG INTERSPERSED NUCLEOTIDE ELEMENTS, and SHORT INTERSPERSED NUCLEOTIDE ELEMENTS) transpose via reverse transcription of an RNA intermediate. Class II elements (or DNA TRANSPOSABLE ELEMENTS—such as transposons, Tn elements, insertion sequence elements, and mobile gene cassettes of bacterial integrons) transpose directly from one site in the DNA to another.

LINEs (Long Interspersed Nuclear Elements or Long INterspersed Elements) Families of long sequences (average length = 6,500 bp), moderately repetitive (about 10,000 copies). LINEs are cDNA copies of functional genes present in the same genome; also known as processed pseudogenes.

Highly repeated sequences, 6K–8K base pairs in length, which contain RNA polymerase II promoters. They also have an open reading frame that is related to the reverse transcriptase of retroviruses, but they do not contain LTRs (long terminal repeats). Copies of the LINE 1 (L1) family form about 15 percent of the human genome. The jockey elements of Drosophila are LINEs.

LTR (Long Terminal Repeat) A sequence directly repeated at both ends of a defined sequence, of the sort typically found in retroviruses.

SINEs (Short Interspersed Nuclear Elements or Short INterspersed Elements) Short interspersed nuclear elements. Families of short (150 to 300 bp), moderately repetitive elements of eukaryotes, occurring about 100,000 times in a genome. SINES appear to be DNA copies of certain tRNA molecules, created presumably by the unintended action of reverse transcriptase during retroviral infection.

Highly repeated sequences, 100–300 bases long, which contain RNA polymerase III promoters. The primate Alu (ALU ELEMENTS) and the rodent B1 SINEs are derived from 7SL RNA, the RNA component of the signal recognition particle. Most other SINEs are derived from tRNAs, including the MIRs (mammalian-wide interspersed repeats).

retroelements Elements that are transcribed into RNA, reverse-transcribed into DNA and then inserted into a new site in the genome. Long terminal repeats (LTRs) similar to those from retroviruses are contained in retrotransposons and retrovirus-like elements. Retroposons, such as LONG INTERSPERSED NUCLEOTIDE ELEMENTS and SHORT INTERSPERSED NUCLEOTIDE ELEMENTS do not contain LTRs.

reverse transcriptase An enzyme, requiring a DNA primer, that catalyzes the synthesis of a DNA strand from an RNA template. An enzyme that can use RNA as a template to synthesize DNA.

ribosome A complex organelle (composed of proteins plus rRNA) that catalyzes translation of messenger RNA into an amino acid sequence. Ribosomes are made up of two nonidentical subunits, each consisting of a different rRNA and a different set of proteins.

RNA (ribonucleic acid) A single-stranded nucleic acid similar to DNA but having ribose sugar rather than deoxyribose sugar and uracil rather than thymine as one of the pyrimidine bases.

messenger RNA (mRNA) An RNA molecule transcribed from the DNA of a gene, and from which a protein is translated by the action of ribosomes. The basic function of the nucleotide sequence of mRNA is to determine the amino acid sequence in proteins.

micro-RNA (miRNA or μRNA) A tiny piece of RNA, about 21 to 23 bases in length, that binds to matching pieces of messenger RNA to make it double-stranded and decrease the production of the corresponding protein. Micro-RNAs were first discovered in the roundworm C. elegans in the early 1990s and are now known in many species, including humans. It is now estimated there are approximately 250–350 micro-RNA molecules in humans. These molecules appear to serve as master regulatory molecules for many biological processes, turning on or off genes coding for proteins.

noncoding RNA (ncRNA) Any RNA molecule with no obvious protein-coding potential for at least 80 or 100 amino acids, as determined by scanning full-length cDNA sequences. It includes ribosomal (rRNA) and transfer RNAs (tRNA) and is now known to include various subclasses of RNA, including snoRNA, siRNA, and piRNA. Just like the coding mRNAs, a large proportion of ncRNAs are transcribed by RNA polymerase II and are large transcripts. A description of the many forms of ncRNA can be found at http://en.wikipedia.org/wiki/Non-coding_RNA.

pre-mRNA The first (primary) transcript from a protein-coding gene is often called a pre-mRNA and contains both introns and exons. Pre-mRNA requires splicing (removal) of introns to produce the final mRNA molecule containing only exons.

ribosomal RNA (rRNA) A class of RNA molecules, coded in the nucleolar organizer, that have an integral (but poorly understood) role in ribosome structure and function. RNA components of the subunits of the ribosomes.

small interfering RNA, or silencing RNA (siRNA) A class of short (20–25·nt), double-stranded RNA molecules. It is involved in the RNA interference pathway, which alters RNA stability and thus affects RNA concentration and thereby suppresses the normal expression of specific genes. Widely used in biomedical research to ablate specific genes.

small nucleolar RNA (snoRNA) A subclass of RNA molecules involved in guiding chemical modification of ribosomal RNA and other RNA genes as part of the regulation of gene expression.

transfer RNA (tRNA) Small RNA molecules that carry amino acids to the ribosome for polymerization into a polypeptide. During translation, the amino acid is inserted into the growing polypeptide chain when the anticodon of the tRNA pairs with a codon on the mRNA being translated.

RNA-induced silencing complex (RISC) A protein complex that mediates the double-stranded RNA-induced destruction of homologous mRNA.

RNA interference or RNA-mediated interference (RNAi) The process by which double-stranded RNA triggers the destruction of homologous mRNA in eukaryotic cells by the RISC.

single nucleotide polymorphism (SNP) The most common form of DNA variation, alterations to a single base. If the SNP is in a gene, it can disrupt the gene's function. Most SNPs do not occur in genes, but can be associated with other types of DNA variation and so are used effectively as markers.

An SNP is a position in the genome where some individuals have one DNA base (e.g., A), and others have a different base (e.g., C). SNPs and point mutations are structurally identical, differing only in their frequency. Variations that occur in 1 percent or less of a population are considered point mutations, and those occurring in more than 1 percent are SNPs. This distinction is pragmatic and reflects the fact that low-frequency mutations cannot be used effectively in genetic studies as genetic markers, while more common ones can.

anonymous SNP SNP occurring in a "junk DNA" region of the genome

cSNP SNP occuring in a coding region of the genome

rSNP SNP occuring in a regulatory region of the genome

spliceosome Protein-RNA complex that removes introns from eukaryotic nuclear RNAs.

splicing The reaction that removes introns and joins together exons in eukaryotic nuclear primary RNA transcripts.

alternative splicing The production of two or more distinct mRNAs from RNA transcripts having the same sequence via differences in splicing (by the choice of different exons).

stochastic process A partially random or uncertain, not continuous set of variables. A stochastic variable is neither completely determined nor completely random; in other words, it contains an element of probability. A system containing one or more stochastic variables is probabilistically determined. In the simplest possible case (see *discrete-time signal*), a stochastic process amounts to a sequence of random variables known as a time series (for example, see *Markov chain*). One approach to stochastic processes treats them as a function of one or several deterministic arguments ("inputs," in most cases regarded as "time") whose values ("outputs") are random variables: nondeterministic (single) quantities that have certain probability distributions (Wikipedia, "Stochastic process").

tandem repeats Multiple copies of the same base sequence on a chromosome; used as a marker in physical mapping.

telomere The ends of linear chromosomes that are required for replication and stability. The tip (or end) of a chromosome.

transcription The synthesis of RNA using a DNA template. The process whereby RNA is synthesized from a DNA template.

translation The process of protein synthesis whereby the primary structure of the protein is determined by the nucleotide sequence in mRNA. The ribosome-mediated production of a polypeptide whose amino acid sequence is derived from the codon sequence of an mRNA molecule.

transposon A mobile piece of DNA that is flanked by terminal repeat sequences and typically bears genes coding for transposition functions. A mobile genetic element that can replicate itself and insert itself into the genome, including interrupting genes and disrupting their function, an insertional mutagen.

One of a class of genes that are capable of moving spontaneously from one chromosome to another, or from one position to another in the same chromosome; also known as jumping genes or transposable elements.

DNA elements carrying genes for transposition and other genetic functions. In many cases, the latter genes enable bacteria to live in extreme environments. Transposons are much longer than IS (Insertion) elements. Abbreviated Tn.

First recognized in the 1940s by Dr. Barbara McClintock in studies of peculiar inheritance patterns found in the colors of Indian corn. Also known as "jumping DNA," referring to the fact that some stretches of DNA are unstable and "transposable," that is, they can move around—on and between chromosomes. This theory was confirmed in the 1980s when scientists observed jumping DNA in other genomes.

retrotransposon DNA fragments copied from viral RNA with reverse transcriptase that insert in the host chromosomes.

virus A particle consisting of a nucleic acid (RNA or DNA) genome surrounded by a protein coat (capsid) and sometimes also a membrane, which can replicate only after infecting a host cell. A virus particle may exist free of its host cell but is incapable of replicating on its own.

retrovirus An RNA virus that replicates by first being converted into double-stranded DNA by reverse transcriptase.

wild-type The most frequently encountered genotype in natural breeding populations. The term "wild-type" was fixed in the lexicon in the early days of fruit-fly genetics, when one could go out and catch one; now it means the original line of normally functioning individuals.

zygote The unique diploid cell formed by the fusion of two haploid cells (often an egg and a sperm) that will divide mitotically to create a differentiated diploid organism.

heterozygote An individual having a heterozygous gene pair. A diploid or polyploid with different alleles at a particular locus.

heterozygous gene pair A gene pair having different alleles in the two chromosome sets of the diploid individual.

Notes

Coda

1. Giorgio Agamben, *The Coming Community*, trans. Michael Hardt (Minneapolis: University of Minnesota Press, 1993). Agamben also wrote that in this age "we are all virtually *homines sacri*" (in *Homo Sacer: Sovereign Power and Bare Life*, trans. Karen Pinkus and Michael Hardt [Minneapolis: University of Minnesota Press, 1998], 115). This book attempts to take seriously this statement, and, following his Greek-inspired distinction, devotes his first part to the zoology of these *homines sacri* (the molecular, or the logics of the *zoe*, i.e., barest life), and his second part to their biology (the molar, or the logics of the *bios*). Junk, then, appears to be the name of their conflation, and *Homo nexus* the eschatological name of this becoming-junk of *Homo sacer*, when everybody has become an instance Homo sacer (and yes, it has to be an endnote).

2. Gilbert Simondon, *L'individuation à la lumière des notions de forme et d'information* (Grenoble: Millon, 2005), 32, translated as "The Genesis of the Individual" in Jonathan Crary and Sanford Kwinter, eds., *Incorporations* (New York: Zone Books, 1994), 297–319. See also Alberto Toscano, *The Theatre of Production: Philosophy and Individuation between Kant and Deleuze* (Houndmills: Palgrave MacMillan, 2006), and especially the fourth section of his fifth chapter, "Transduction: Search for a Method": "For Simondon, the Parmenidean unity of being and thought is a matter of contagion rather than totalization; a matter of ontogenesis rather than critique" (151). Much of the present book focuses on an attempt to open this ontogenesis toward a critique.

3. By the way, "transductive" also refers to a specific form of reasoning qualifying an infant stage of development: according to Piaget, it defines "the primary form of reasoning used during preoperational stage of development, from ages two to seven. Its most basic form is "if A causes B today, then A always causes B" (Wikipedia). So be it.

Introduction

1. Thomas Pynchon, *Gravity's Rainbow* (New York and London: Viking/Jonathan Cape, 1973), 18. See Harjo Berressem, *Pynchon's Poetics* (Champaign: University of Illinois Press, 1993), 7, for the analogy with Foucault.

2. According to some recent figures, cars are the most recycled products in the

United States, where 95 percent of them are recycled. See *The Car Connection Blog*, http://www.thecarconnection.com/blog/?p=1159, accessed April 28, 2008.

3. In *Life and Habit* (Trübner, 1878), 134.

4. In *The Notebooks of Samuel Butler* (London: Hogarth Press, 1985), 44–46.

5. Ibid., 21.

6. Ibid., 95.

7. Michel Serres, *Genesis*, trans. Geneviève James and James Nielson (Ann Arbor: University of Michigan Press, 1995), 32.

8. http://www.bartleby.com/61/35/J0083500.html.

9. http://www.etymonline.com.

10. Merriam-Webster's online entry "rush," http://www.m-w.com/cgi-bin/dictio nary?book=Dictionary&va=rush.

11. http://www.bootlegbooks.com/Reference/PhraseAndFable/data/693.html.

12. http://www.geocities.com/indoeurop/project/phonetics/word58.html.

13. http://members.iinet.net.au/~weeds/western_weeds/juncaceae_orchid aceae.htm.

14. Gilles Deleuze and Félix Guattari, *A Thousand Plateaus: Capitalism and Schizophrenia*, trans. Brian Massumi (Minneapolis: University of Minnesota Press, 1987), 25.

15. Ibid., 7–13.

16. For a contextualization of Deleuze and Guattari's biophilosophy with respect to molecular biology, see Mark Hansen, "Becoming as Creative Evolution?" *Postmodern Culture* 11(1) (2000), and John Marks, "Molecular Biology in the Work of Deleuze and Guattari," *Paragraph* 29(2) (2006): 81–87.

17. Francis H. C. Crick and James D. Watson, "Molecular Structure of Nucleic Acids: A Structure for Deoxyribose Nucleic Acid," *Nature* 171(4356) (April 25, 1953): 737.

18. Francis H. C. Crick, "On Protein Synthesis," *Symposium of the Society of Experimental Biology* 12: 138–63 (New York: Academic Press, 1958).

19. Human Genome Project Information Web site, http://www.ornl.gov/sci/tech resources/Human_Genome/faq/faqs1.shtml, accessed August 11, 2007.

20. Algirdas Julien Greimas, "Éléments d'une grammaire narrative," in *Du sens: Essais sémiotiques* (Paris: Seuil, 1970).

21. David Bloor, *Knowledge and Social Imagery*, 2d ed. (Chicago: Chicago University Press, 1991 [1976 for the first edition]), 5–8.

22. With such an understanding of the "object" of this book, I firmly locate my endeavor in the critical footprints left by Walter Benjamin: "Benjamin's research," writes Giorgio Agamben, "favors these objects, which in this very semblance of being secondary, or even destined to scrap (Benjamin talks about the 'rags' of history), present with more strength a kind of signature or mark that sends them back to the present" (*Signatura rerum: Sur la méthode* [Paris: Vrin, 2008], 83, my translation).

23. Michel Foucault, "L'éthique du souci de soi comme pratique de la liberté," in *Dits et écrits* (1984) (Paris: Gallimard, 2001), 1530–31, my translation.

24. Or, to put it more bluntly, "*capitalism itself has become Deleuzian in form, in style,*

and in content" (Frédéric Vandenberghe, "Deleuzian capitalism," *Philosophy and Social Criticism* 34[8] [2008]: 879).

25. Guy Debord, *La société du spectacle* (Paris: Gallimard [Folio ed.], 1992 [1967]), 19, my translation.

26. Jean Baudrillard, *Simulacra and Simulation*, trans. Sheila Faria Glaser (Ann Arbor: University of Michigan Press, 1994 [1981]), 1.

27. Hillel Schwartz, *The Culture of the Copy: Striking Likenesses, Unreasonable Facsimiles* (New York: Zone Books, 1996), 11.

1. How Junk Became, and Why It Might Remain, Selfish

1. David Baltimore, "Our Genome Unveiled," *Nature* 409 (February 15, 2001): 814–16.

2. "Discovery commences with the awareness of anomaly, i.e., with the recognition that nature has somehow violated the paradigm-induced expectations that govern normal science" (Thomas Kuhn, *The Structure of Scientific Revolutions* [Chicago: University of Chicago Press, 1962], 52).

3. Paul Berg, "All Our Collective Ingenuity Will Be Needed," *FASEB Journal* 5(75) (1991).

4. J. C. Venter et al., "The Sequence of the Human Genome," *Science* 291(5507) (February 16, 2001): 1304–51.

5. Along with Renato Dulbecco and Howard Martin Temin, David Baltimore won the Nobel Prize in medicine in 1975 "for their discoveries concerning the interaction between tumor viruses and the genetic material of the cell." Baltimore's work centered on the characterization of the reverse-transcriptase, a specific enzyme in RNA tumor virus particles that could make a DNA copy from RNA.

6. Baltimore, "Our Genome Unveiled," 815.

7. Berg, "All Our Collective Ingenuity Will Be Needed."

8. Francis Crick, "Split Genes and RNA Splicing," *Science* 204 (1979): 264 and 270.

9. Susumu Ohno, *Evolution by Gene Duplication* (New York: Springer-Verlag, 1970), 18. The reference is Roy J. Britten and David E. Kohne, "Repeated Sequences in DNA," *Science* 161 (1968): 529–40. I shall come back to Ohno's work in the following chapter.

10. Wen Hsiung Li, *Molecular Evolution* (Sunderland, Mass.: Sinauer Associates, 1997). The references are Emil Zuckerkandl, "Gene Control in Eukaryotes and the C-Value Paradox: 'Excess' DNA as an Impediment to Transcription of Coding Sequences," *Journal of Molecular Evolution* 9 (1976): 73–104; Susumu Ohno, "So Much 'Junk' DNA in Our Genome," in *Evolution of Genetic Systems*, ed. H. H. Smith (New York: Gordon and Breach, 1972), 366–70; G. Östergren, "Parasitic Nature of Extra Fragment Chromosomes," *Botaniska Notiser* 2 (1945): 157–63; Thomas Cavalier-Smith, "Nuclear Volume Control by Nucleoskeletal DNA, Selection for Cell Volume and Cell Growth Rate, and the Solution of the DNA C-Value Paradox," *Journal of Cell Science* 34 (1978): 247–78.

11. A glossary provided at the end of this volume offers standard definitions of these technical terms.

12. Thomas Cavalier-Smith, "Introduction: The Evolutionary Significance of Genome Size," in *The Evolution of Genome Size*, ed. Thomas Cavalier-Smith (Chichester, U.K.: John Wiley & Sons, 1985), 1–2. The references are A. E. Mirsky and H. Ris, "The Deoxyribonucleic Acid Content of Animal Cells and Its Evolutionary Significance," *Journal of General Physiology* 34 (1951): 451–62, and C. A. Thomas, "The Genetic Organization of Chromosomes," *Annual Review of Genetics* 5 (1971): 237–56.

13. T. Ryan Gregory, "The C-Value Enigma," PhD thesis, Department of Zoology, University of Guelph, Guelph, Ontario, Canada, 2002, 7.

14. Ibid., 3.

15. See Richard Doyle, *On Beyond Living* (Stanford, Calif.: Stanford University Press, 1997); Lily Kay, *Who Wrote the Book of Life? A History of the Genetic Code* (Stanford, Calif.: Stanford University Press, 2000); Evelyn Fox Keller, *Making Sense of Life: Explaining Biological Development with Models, Metaphors and Machines* (Cambridge: Harvard University Press, 2002); Judith Roof, *The Poetics of DNA* (Minneapolis: University of Minnesota Press, 2007).

16. Werner R. Loewenstein, *The Touchstone of Life: Molecular Information, Cell Communication and the Foundations of Life* (Oxford: Oxford University Press, 1999), 93.

17. Nathalie Angier, "Biologists Seek Words in DNA's Unbroken Text," *New York Times*, July 9, 1991, C1.

18. Umberto Eco, *A Theory of Semiotics* (Bloomington: Indiana University Press, 1976), 36–38.

19. Or "Structural code."

20. Roland Barthes, "L'effet de réel," *Communications* 11 (1968): 88, my translation.

21. Ibid., 87.

22. Jean-Pierre Faye, *Le récit hunique* (Paris: Seuil, 1967).

23. Jean-Pierre Faye, *Le langage meurtrier* (Paris: Hermann, 1996).

24. Ibid., 2, my translation.

25. Jorge Luis Borges, "L'écriture du Dieu," in *Œuvres Complètes*, vol. 1, Collection La Pléiade (Paris: Gallimard, 1949), 633, my translation.

26. McFadden's first chapter is titled "What Is Life?" just like Schrödinger's book.

27. Johnjoe McFadden, *Quantum Evolution: The New Science of Life* (New York: Norton, 2000), 268.

28. Richard Dawkins, *The Selfish Gene* (London and New York: Norton, 1989 [1976]), 44–45.

29. Richard Dawkins, *The Blind Watchmaker* (New York: Norton, 1987), 116.

30. Ibid., 174.

31. Ibid.

32. Ibid.

33. Richard Dawkins, "In Defense of Selfish Genes," available online at http://www.royalinstitutephilosophy.org/articles/dawkins_genes.htm.

34. See Thierry Bardini, *Bootstrapping: Douglas Engelbart, Coevolution and the Origins of Personal Computing* (Stanford, Calif.: Stanford University Press, 2000), 43–45.

35. Jeremy Narby, *The Cosmic Serpent: DNA and the Origins of Knowledge* (New York: Jeremy P. Tarcher/Putnam, 1998).

36. Ibid., 11.

37. Ibid., 34.

38. Ibid., 71.

39. Ibid., 145.

40. W. Ford Doolittle and Carmen Sapienza, "Selfish Genes, the Phenotype Paradigm and Genome Evolution"; Leslie E. Orgel and Francis H. C. Crick, "Selfish DNA: The Ultimate Parasite," *Nature* 284 (1980): 601–3; 604–7.

41. Doolittle and Sapienza, "Selfish Genes," 603.

42. Orgel and Crick, "Selfish DNA," 607.

43. Thomas Cavalier-Smith, "How Selfish Is DNA?"; R. A. Reid, "Selfish DNA and 'Petite' Mutants"; Gabriel Dover, "Ignorant DNA?"; Temple F. Smith, "Occam's Razor," *Nature* 285 (1980): 617–18; 618; 618–20; 620.

44. Cavalier-Smith, "How Selfish Is DNA?" 617.

45. Dover, "Ignorant DNA?" 618–19.

46. Leslie E. Orgel, Francis H. C. Crick, and Carmen Sapienza, "Selfish DNA"; Gabriel Dover and W. Ford Doolittle, "Modes of Genome Evolution," *Nature* 288 (1980): 645–46; 646–47.

47. Orgel, Crick, and Sapienza, "Selfish DNA," 646.

48. Dover and Doolittle, "Modes of Genome Evolution," 646.

49. Ibid., 647.

50. Orgel, Crick, and Sapienza, "Selfish DNA," 645.

51. Available at the Welcome Library for the History and Understanding of Medicine, File PP/CRI H/6/13/4. The 92 exposures of the file include 9 letters and 7 manuscripts of the papers leading to the two final replies published. Here is a detailed table of contents of the correspondence used hereafter: 08/06/1980 FDtoLO (Doolittle to Orgel) Ms.0 and Ms.1.1; 10/23/1980 GDtoFC (Dover to Crick); 10/30/1980 LOtoFC (Orgel to Crick); 11/05/1980 FDtoPN (Doolittle to Newmark) final Ms.1; 11/06/1980 FDtoFC (Doolittle to Crick) final Ms.1; 11/11/1980 FCtoCS (Crick to Sapienza) Ms.2.2; 11/11/1980 FCtoGD (Crick to Dover) Ms.2.1; 11/20/1980 FCtoPN (Crick to Newmark) final Ms. 2; 11/20/1980 FCtoFD (Crick to Doolittle) final Ms. 2.

52. Reid, "Selfish DNA and 'Petite' Mutants," 620.

53. Smith, "Occam's Razor," 620.

54. Charles Sanders Peirce, *Collected Papers of Charles Sanders Peirce*, ed. Charles Hartshorne and Paul Weiss (Cambridge: Belknap Press of Harvard University Press, 1932), 4:1.

55. Paul Vincent Spade, "Ockham's Nominalist Metaphysics: Some Main Themes," in *The Cambridge Companion to Ockham*, ed. Paul Vincent Spade (Cambridge: Cambridge University Press, 1999), 102.

56. Ibid., 101–2.

57. Smith, "Occam's Razor," 620.

58. Stephen Jay Gould, "The Ultimate Parasite," *Natural History* 90 (1981): 10. One is reminded here of Karl Popper's famous "recantation": "I have in the past described the theory as "almost tautological," and I have tried to explain how the theory of natural selection could be untestable (as is a tautology) and yet of great scientific interest" (Karl R. Popper, "Natural Selection and the Emergence of Mind," *Dialectica* 32[3–4] [1978]: 344–46).

59. Gould, "The Ultimate Parasite," 11.

60. Alain de Libera, *La querelle des universaux: De Platon à la fin du Moyen Âge* (Paris: Seuil, 1996).

61. A starting point would be the extensive discussion of the hierarchical theory of selection provided by Gould in his final opus, *The Structure of Evolutionary Theory* (Cambridge: Belknap Press of Harvard University Press, 2002). Gould confesses that he "has struggled with this issue [how to define an "individual"] all [his] professional life, and [has] often wondered why the questions raised seem so much more recalcitrant, and so much more cascading in implications, than for any other major problem in Darwinian theory" (598). Gould developed arguments very close to those I am presenting here: (1) he diagnosed a problematic use of Occam's razor in the doctrine leading to the selfish gene hypothesis: more reductionism than parsimony in his eye (550–56), and (2) he showed that this doctrine commits a logical fallacy in its attempt to establish the gene as the sole locus of selection; he characterized this "fruitful error of logic" (613) as "a confusion of bookkeeping with causality" (643–44).

62. Orgel, Crick, and Sapienza, "Selfish DNA," 645.

63. Cavalier-Smith, "How Selfish Is DNA?" 618.

64. Richard Dawkins, *The Extended Phenotype* (Oxford: Oxford University Press, 1999 [1982]). See especially his chapter 9 titled "Selfish DNA, Jumping Genes and a Lamarckian Scare" (156–78), where Dawkins directly refers to the controversy studied here and offers his appraisal of the two review essays that started it: "It is because they are starting to do just this [turn the same cynical eye on the cell's 'own' DNA] that I find the papers of Doolittle and Sapienza and Orgel and Crick so exciting, in comparison with objections of Cavalier-Smith (1980), Dover (1980) and others, although of course the objectors may be right in the particular points they make" (163).

65. Gould, *The Structure of Evolutionary Theory*, 638.

66. Steven Shapin, *A Social History of Truth: Civility and Science in Seventeenth-Century England* (Chicago: University of Chicago Press, 1994), 68.

67. Ibid., 191.

68. See Roy J. Britten and Eric H. Davidson, "Gene Regulation for Higher Cells: A Theory," *Science* 165 (1969): 349–57; Roy J. Britten and Eric H. Davidson, "Repetitive and Non-repetitive DNA Sequences and a Speculation on the Origins of Evolutionary Novelty," *Quarterly Review of Biology* 46 (1971): 111–38.

69. Roy J. Britten, personal interview with the author, Corona del Mar, California, October 12, 2003.

70. In his comments on the present chapter, Roy Britten wrote: "It seems to me you

have made a basic error in connecting Dawkins to the other parts of this issue. As I understand he considers genes to be selfish—real functioning genes just help the animals out because it is good for the genes themselves. In contrast selfish DNA in the Doolittle sense is nonfunctional and exists only because it is selfish, i.e., 'has learned how to get itself copied.' This personal phrasing 'itself' and 'selfish' is nonsense. Even parasitic is a little dubious" (personal communication, November 9, 2004). It is precisely my point in this chapter that this connection is problematic: "selfish gene" and "selfish DNA" are two connected issues that reflect the basic problem of the two contradictory definitions of "genome" presented at the beginning of this chapter. Selfish *genes* and selfish *DNA* were used in the two titles of the 1980 issue of *Nature* studied here.

Britten added: "It is now popular to believe that all of the extra DNA is from mobile elements—50% of the human genome—and they are not little persons capable of considering their own and their copies' selfish interests. Their copying is definitely a molecular-mechanical mechanism. Now do they do the organism some good? You know I have published several papers showing that they play regulatory roles. I am now publishing a paper showing that a half dozen of them have been converted *in toto* to functioning human genes. Many fragments of gene coding sequences derive from them. They are definitely useful—but we only know a small extent of usefulness. They may be useful on a grand scale and we still have to learn in what way. Their existence could be because of their usefulness. Or their usefulness could be because of their existence. Living systems will take advantage of whatever is around." The reference is to Roy J. Britten, "Coding Sequences of Functioning Human Genes Derived Entirely from Mobile Element Sequences," *Proceedings of the National Academy of Sciences* 101(48) (2004): 16825–30.

71. Evelyn Fox Keller, *A Feeling for the Organism: The Life and Work of Barbara McClintock* (New York: W. H. Freeman, 1983).

72. Ibid., xiii.

73. Ibid., 207.

74. Jacques Vallée and Janine Vallée, *Challenge to Science: The UFO Enigma* (Chicago: Regnery, 1966).

75. Colm A. Kelleher, personal interview with the author, Las Vegas, October 10, 2003.

76. Sidney Brenner, "Refuge of Spandrels," *Current Biology* (1999): 669, quoted in Gould, *The Structure of Evolutionary Theory*, 1269.

77. Keller, *Making Sense of Life*, 131.

78. Ibid.

79. Lenny Moss, *What Genes Can't Do*, first paperback edition (Cambridge: MIT Press, 2004), back cover.

80. Ibid., 184.

81. Thierry Bardini, "Variations on Genetic Insignificant: Metaphors of the (non)-Code," in *The Tasking of Identity in Contemporary Art: At the Intersection of Aesthetics,*

Media, Science and Technology, ed. Christine Ross, Johanne Lamoureux, and Olivier Asselin (Montreal: McGill/Queens University Press, 2008), 433–51.

82. Moss, *What Genes Can't Do*, 185. The reference is to Scott F. Gilbert, "The Embryological Origins of the Gene Theory," *Journal of the History of Biology* 11 (1978): 307–51.

83. Ibid., 194.

84. Jesper Hoffmeyer and Claus Emmeche, "Code-Duality and the Semiotics of Nature," in *On Semiotic Modeling*, ed. M. Anderson and F. Merrell (Berlin and New York: Mouton de Gruyter, 1991), 117–66.

85. Wikipedia: http://en.wikipedia.org/wiki/Junk_DNA, accessed September 25, 2005. As of August 19, 2007, the first paragraph has been corrected as follows: "About 80–90% of the human genome has been designated as 'junk,' including most sequences within introns and most intergenic DNA. While much of this sequence may be an evolutionary artifact that serves no present-day purpose, some is believed to function in ways that are not currently understood. Moreover, the conservation of some junk DNA over many millions of years of evolution may imply an essential function. Some consider the 'junk' label as something of a misnomer, but others consider it apposite as junk is stored away for possible new uses, rather than thrown out; others prefer the term 'noncoding DNA' (although junk DNA often includes transposons that encode proteins with no clear value to their host genome). However, it now appears that, although protein-coding DNA makes up barely 2% of the human genome, about 80% of the bases in the genome may be being expressed, which supports the view that the term 'junk DNA' may be a misnomer."

86. EvoWiki http://www.evowiki.org/index.php/Junk_DNA, accessed September 25, 2005.

87. Ibid.

88. One of the earliest enunciations of an actual relationship between a possible function for junk DNA and a neocreationist stand can be found in a paper by Jerry Bergman titled "The Functions of Introns: From Junk DNA to Designed DNA," *Perspectives on Science and Christian Faith* 53(3) (September 2001). Bergman concluded: "In the past, evolutionary geneticists, once uncertain as to what this apparently superfluous DNA does, referred to introns and other non-coding DNA as 'junk.' Evidence is now being accumulated which indicates that much or most of this DNA may not be junk, but critical for life itself. If functions for most or all of the non-coding DNA is found, Darwinism would be without the raw material needed to produce new genes by mutations that can be selected for evolution to occur. Furthermore, much of this new information on the complexity of the genome elegantly provides evidence for both intelligent design and for the concept of irreducible complexity." See http://www.rae.org/introns.html, accessed September 25, 2005.

89. The full paper is available on the Institute's Web page: http://www.discovery.org, accessed September 26, 2005.

90. Among the many online references on this controversy: Dr. Richard Sternberg's homepage: http://www.rsternberg.net: "Meyer's hopeless monster": A negative

review of Meyer's paper by Wesley R. Elsberry on the evolutionist Web site "The Panda's Thumb": http://www.pandasthumb.org/pt-archives/000430.html: "Unintelligent design," a description of the controversy by David Klinghoffer for the *National Review Online:* http://www.nationalreview.com/comment/klinghoffer200508160826.asp: "Editor Explains Reasons for 'Intelligent Design' Article" by Michael Powell for the *Washington Post* online: http://www.washingtonpost.com/wpdyn/content/article/2005/08/18/AR2005081801680_pf.html.

91. Ron Hutcheson, "Bush Endorses Teaching 'Intelligent Design' Theory in Schools," Knight Ridder Newspapers, online at http://www.realcities.com/mld/krwashington/12278497.htm.

92. Susan Jacoby, "Caught between Church and State," *New York Times*, January 19, 2005, http://www.theocracywatch.org/schools_jacoby_times_jan19_05.htm, accessed September 27, 2005.

93. Neil A. Manson, "Anthropocentrism and the Design Argument," *Religious Studies* 36 (2000): 163–76.

94. And especially in the philosophies of Anaximander, Heraclitus, and Anaxagoras. See Frederik Ferre's "Design Argument," in the *Dictionary of the History of Ideas: Studies of Selected Pivotal Ideas*, vol. 1, ed. Philip P. Wiener (New York: Charles Scribner's Sons, 1973–74), 670–77, online at http://etext.lib.virginia.edu/cgilocal/DHI/dhiana.cgi?id=dv1-80, accessed October 1, 2005.

95. Ibid., 672–73.

96. "Through the seventeenth century it continued to be the scientists, or those with deep scientific interests, who stated the design argument with most force" (ibid., 674).

97. And, not long later, as English latitudinarianism held, via the Book of Nature only.

98. Ferre, "Design Argument," 674.

99. Steven Shapin and Simon Schaffer, *Leviathan and the Air-Pump: Hobbes, Boyle, and the Experimental Life* (Princeton, N.J.: Princeton University Press, 1989), 24.

100. http://en.wikipedia.org/wiki/Bernard_Nieuwentyt.

101. William Paley, *Natural Theology: or, Evidences of the Existence and Attributes of the Deity, Collected from the Appearances of Nature*, 12th ed. (London: J. Faulder, 1809), 1–3, available online at http://www.hti.umich.edu/cgi/p/pd-modeng/pd-modeng-idx?type=HTML&rgn=TEI.2&byte=53049319, accessed November 23, 2006.

102. Fernando Vidal, "Extraordinary Bodies and the Physicotheological Imagination," in *The Faces of Nature in Enlightenment Europe*, ed. Gianna Pomata and Lorraine Daston (Berlin: Berliner-Wissenschafts-Verlag, 2003), 61–96.

103. In Nora Barlow, ed., *The Autobiography of Charles Darwin, 1809–1882: With Original Omissions Restored* (New York: Norton, 1969 [1958]), 87.

104. See Bruce Mazlish, *The Fourth Discontinuity: The Co-Evolution of Humans and Machines* (New Haven: Yale University Press, 1993).

105. Asa Gray, "Natural Selection Not Inconsistent with Natural Theology," *Atlantic*

Monthly (October 1860), reprinted in *Darwiniana: Essays and Reviews Pertaining to Darwinism* (New York 1876), electronic version available thanks to the Gutenberg Project, http://www.gutenberg.org/etext/5273.

106. Dov Ospovat, "God and Natural Selection: The Darwinian Idea of Design," *Journal of the History of Biology* 13(169–74) (1980): 187.

107. Ibid., 188.

108. In a letter to Asa Gray written in 1860 and quoted in ibid., 189.

109. Francis Darwin, ed., *The Life and Letters of Charles Darwin* (New York: Basic Books, 1959), 395.

110. Gray, "Natural Selection Not Inconsistent with Natural Theology."

111. William A. Dembski, "Intelligent Design: Yesterday's Orthodoxy, Today's Heresy," edited transcript from a lecture given Saturday, January 17, 2004, at Grace Valley Christian Center, Davis, California, as part of the Faith and Reason series sponsored by Grace Alive! and Grace Valley Christian Center, http://www.designinference.com/documents/2005.04.ID_Orthodoxy_Heresy.htm.

112. C. G. Hunter, *Darwin's Proof: The Triumph of Religion over Science* (Grand Rapids, Mich.: Brazos Press, 2003), 118.

113. Patrick Edward Dove, *The Theory of Human Progression and Natural Probability of Justice*, 476–77, available online at Google Books (http://books.google.com), accessed December 19, 2006.

114. Elliott Sober, "The Design Argument," updated from his published version in W. Mann, ed., *The Blackwell Companion to Philosophy of Religion* (2004), available online at http://philosophy.wisc.edu/sober/design%20argument%2011%202004.pdf, accessed October 1, 2005.

2. From Garbage to Junk DNA

1. "'Waste became an essential component of the production cycle' because a market would otherwise fail when it reached saturation point and thus 'industry . . . needed to shorten the natural "lifespan" of "durable" products so that people would buy them not once but several times, thereby stabilizing the production cycle'" (John Scanlan quoting Ellen Lupton and J. A. Miller, *On Garbage* [London: Reaktion Books, 2005], 38).

2. Barry Allen, "The Ethical Artifact: on Trash," in *Trash*, ed. John Knechtel (Cambridge: MIT Press, 2007), 202.

3. Bruno Mouron and Pascal Rostain, *Trash*, exhibited at La Maison Européenne de la Photographie, Paris, March 14–June 3, 2007, http://www.mep-fr.org/actu_1.htm.

4. Allen, "The Ethical Artifact," 202.

5. Gay Hawkins, "Sad Chairs," in Mouron and Rostain, *Trash*, 55.

6. Ibid.

7. Scanlan, *On Garbage*, 22.

8. Ibid., 37.

9. Susumu Ohno, *Evolution by Gene Duplication* (New York: Springer-Verlag, 1970), n.p.

10. Ernest Beutler, *Susumu Ohno, 1928–2000: A Biographical Memoir,* National Academy of Science Biographical Memoirs 81 (Washington, D.C.: National Academies Press, 2002), 8.

11. Elizabeth Pennisi, "Genome Duplications: The Stuff of Evolution?" *Science* 294(5551) (December 21, 2001): 2458–60.

12. Ibid.

13. Beutler, *Susumu Ohno,* 6–7.

14. Ohno, *Evolution by Gene Duplication,* 26.

15. Ibid., 62.

16. Ibid.

17. Susumu Ohno, "So Much 'Junk' DNA in Our Genome," in *Evolution of Genetic Systems,* ed. H. H. Smith, (New York: Gordon and Breach, 1972), 368.

18. Susumu Ohno, "An Argument for the Genetic Simplicity of Man and Other Mammals," *Journal of Human Evolution* 1 (1972): 651–62.

19. Ohno, "So Much 'Junk' DNA in Our Genome," 366–67; references elided.

20. T. Ryan Gregory, "A Word about Junk DNA," in *Genomicron: Exploring Genomic Diversity and Evolution,* April 11, 2007, http://genomicron.blogspot.com/2007/04/word-about-junk-dna.html, accessed May 18, 2007.

21. Ohno, "So Much 'Junk' DNA in Our Genome," 367–68.

22. Ibid., 369.

23. Scanlan, *On Garbage,* 36.

24. Ibid., 41.

25. Gregory, "A Word about Junk DNA," references elided.

26. For example, Mark Gerstein and Deyou Zheng, "The Real Life of Pseudogenes," *Scientific American* (August 2006): 48: "Our genetic closet holds skeletons. The bones of long-dead genes—known as pseudogenes—litter our chromosomes."

27. C. Jacq, J. R. Miller, and G. G. Brownlee, "A Pseudogene Structure in 5S DNA of *Xenopus laevis,*" *Cell* 12 (1977): 109–20.

28. William H. Calvin, *The River That Flows Uphill: A Journey from the Big Bang to the Big Brain* (Sierra Club Books, 1987), http://williamcalvin.com/bk3/bk3day6.htm, accessed May 20, 2007.

29. Douglas Hofstadter, *I Am a Strange Loop* (New York: Basic Books, 2007).

30. André-Marie Ampère, *Essai sur la philosophie des sciences* (Paris: Bachelier, 1834).

31. Julien Offray de La Mettrie, *Man a Machine* (La Salle, Ill.: Open Court, 1991), 93.

32. Lewis Mumford, *Technics and Civilization* (New York: Harcourt, 1934), 14.

33. But it can also be argued that it was there since the dawn of the technological ages, with knitting: "The most distinctive feature of knitting is its loops," writes Sadie Plant in "Mobile Knitting," in *Information Is Alive* (Rotterdam: V2_/NAi Publishers, 2003), 30.

34. Norbert Wiener, *Cybernetics, or Control and Communication in the Animal and the Machine,* 2d ed. (Cambridge: MIT Press, 1994 [1948]), 6–7.

35. Pierre de Latil, *La Pensée artificielle: Introduction à la cybernétique* (Paris: Gallimard 1953), quoted in D. J. Stewart, "An Essay on the Origins of Cybernetics," 2000 (1959), http://www.hfr.org.uk/cybernetics-pages/origins.htm, accessed May 26, 2007.

36. *Merriam Webster's Collegiate Dictionary,* 10th ed.

37. Aristotle, *Physics,* II (Sioux Falls, S.Dak.: NuVision Publications, LLC, 2007), 3.

38. Amos Funkenstein, *Theology and the Scientific Imagination from the Middle Ages to the Seventeenth Century* (Princeton, N.J.: Princeton University Press, 1989), 39.

39. Arturo Rosenblueth, Norbert Wiener, and Julian Bigelow, "Behavior, Purpose and Teleology," *Philosophy of Science* 10 (1943): 23.

40. Aristotle, *Physics,* 28.

41. Warren Weaver, in a letter to Norbert Wiener dated January 28, 1949, quoted in Lily Kay, *Who Wrote the Book of Life? A History of the Genetic Code* (Stanford, Calif.: Stanford University Press, 2000), 83.

42. Ibid.

43. Colin S. Pittendrigh, "Temporal Organization: Reflections of a Darwinian Clock-Watcher," *Annual Review of Physiology* 55 (1993): 20. In this paragraph, Pittendrigh alludes to his chapter published in 1958 and titled "Adaptation, Natural Selection, and Behavior," in *Behavior and Evolution,* ed. A. Roe and G. G. Simpson (New Haven: Yale University Press, 1958), 390–416.

44. François Jacob, *The Logic of Life: A History of Heredity,* trans. Betty E. Spillmann (New York: Pantheon Books, 1973 [1970]), 8–9.

45. Richard Dawkins, *The Extended Phenotype: The Long Reach of the Gene,* rev. ed. (Oxford: Oxford University Press, 1999 [1982]), 301–2. Dawkins adds: "This book is in an essay in teleonomy."

46. Ernst Mayr, "Cause and Effect: Kinds of Causes, Predictability, and Teleology Are Viewed by a Practicing Biologist," *Science* 134 (November 1961): 1504.

47. See Ernst von Glasersfeld, "Teleology and the Concepts of Causation," *Philosophica* 46(2) (1990): 17–43. Von Glasersfeld used here Gordon Pask's distinction between *purpose for* and *purpose of* to make this point.

48. Pittendrigh, "Temporal Organization," 20.

49. See Kay, *Who Wrote the Book of Life?,* and especially her chapter 5, "The Pasteur Connection: Cybernétique Enzymatique, Gène Informateur, and Messenger RNA," 193–234. See also Evelyn Fox Keller, *Making Sense of Life: Explaining Biological Development with Models, Metaphors and Machines* (Cambridge: Harvard University Press, 2002), and especially her chapters 4, "Genes, Gene Action, and Genetic Program," 123–47, and 5, "Taming the Cybernetic Metaphor," 148–72.

50. Keller, *Making Sense of Life,* 136–37.

51. Ibid., 170.

52. Ibid., 138–39.

53. Ibid., 146–47.

54. Personal interview with Roy Britten, October 13, 2003.

55. Ibid.

56. Richard Brooke Roberts (1910–80) was another physicist member of the Manhattan project. He joined the Carnegie Institution of Washington in 1937 as its first postdoc in physics and witnessed the first demonstration of nuclear fission in 1939 with prestigious colleagues such Enrico Fermi and Niels Bohr. He did his graduate studies at Princeton too between 1932 and 1936, and this Princeton connection helped Britten get hired in his group in 1951. By then, Roberts had moved to studies in biophysics, "quantitative research in biology carried out by investigators trained in physics" (*CIW Year Book* 50 [1951]). Roberts remains famous for having proposed first the word *ribosome* in 1958, and proved, with his group, that protein synthesis occurred on ribosomes. See Roy J. Britten, "Richard Brooke Roberts," in *Biographical Memoirs*, vol. 62 (Washington, D.C.: National Academy of Sciences Press, 1993), 326–49.

57. R. B. Roberts, P. H. Abelson, D. B. Cowie, E. B. Bolton, and J. R. Britten, *Studies of Biosynthesis in Escherichia coli* (Washington, D.C.: Carnegie Institution of Washington, 1955).

58. Pierre Baldi and G. Wesley Hatfield, *DNA Microarrays and Gene Expression: From Experiments to Data Analysis and Modeling* (Cambridge: Cambridge University Press, 2002).

59. Britten, "Richard Brooke Roberts," 340.

60. See the entry "Roy John Britten" in *World of Genetics* (Thomson Gale, 2005–6), http://www.bookrags.com/biography/roy-john-britten-wog, accessed June 1, 2007.

61. E. T. Bolton et al., *CIW Year Book* 62 (1963): 303, and 63 (1964): 366.

62. B. H. Hover, E. T. Bolton, B. J. McCarthy, and R. B. Roberts, *CIW Year Book* 63 (1964): 394.

63. R. J. Britten and M. Waring, *CIW Year Book* 64 (1965): 316.

64. M. Waring and R. J. Britten, *Science* 154 (1966): 791; R. J. Britten and D. E. Kohne, *CIW Year Book* 65 (1966): 73, and 66 (1967): 73; R. J. Britten and D. E. Kohne, "Repeated Sequences in DNA," *Science* 161 (1968): 528–40.

65. Britten and Kohne, "Repeated Sequences in DNA," 539.

66. "On some early occasions it was suggested that RNA may act as a regulatory molecule. The possibility was first mooted briefly by Jacob and Monod in 1961 . . . but lapsed when the archetypal gene regulatory factor, the *lac* repressor, was subsequently shown to be a protein" (John S. Mattick, "A New Paradigm for Developmental Biology," *Journal of Experimental Biology* 210 [2007]: 1526–47, references elided).

67. Roy J. Britten and Eric H. Davidson, "Gene Regulation for Higher Cells: A Theory," *Science* 165 (1969): 349–57.

68. Britten, personal interview with the author.

69. Bruce S. McEwen and Caleb Finch, "Alfred E. Mirsky and the Foundations of Molecular Biology and Neuroendocrinology," *Endocrinology* 130(1) (1992): 6–7.

70. Seymour S. Cohen, "Alfred Ezra Mirsky," *Biographical Memoirs* 73 (Washington, D.C.: National Academy of Sciences Press, 1998), 322–32.

71. Eric H. Davidson, "Hormones and Genes," *Scientific American* 212 (1965): 36–45.

72. Britten, personal interview with the author.

73. Britten and Davidson, "Gene Regulation for Higher Cells," 349.

74. Ibid., 356.

75. Ibid., 354.

76. Ibid., 353. In this section, the authors quote two recent papers by McClintock: *Brookhaven Symp. Biol.* 18 (1965): 162, and *Develop. Biol. (Suppl.)* 1 (1967): 84.

77. Britten, personal interview with the author.

78. Roy J. Britten, "Coding Sequences of Functioning Human Genes Derived Entirely from Mobile Element Sequences," *Proceedings of the National Academy of Sciences* 101 (2004): 16825–30, and Roy J. Britten, "Transposable Elements Have Contributed to Thousands of Human Proteins," *Proceedings of the National Academy of Sciences* 103 (2006): 1798–1803.

79. Roy J. Britten and Eric H. Davidson, "Repetitive and Non-repetitive DNA Sequences and a Speculation on the Origins of Evolutionary Novelty," *Quarterly Review of Biology* 46 (1971): 111–38; Eric H. Davidson and Roy J. Britten, "Regulation of Gene Expression: Possible Role of Repetitive Sequences," *Science* 204 (1979): 1052–59; G. P. Moore, F. D. Constantini, J. W. Posakony, E. H. Davidson, and R. J. Britten, "Evolutionary Conservation of Repetitive Sequence Expression in Sea Urchin Egg RNA's," *Science* 208 (1980): 1046–48; R. H. Scheller, D. M. Anderson, J. W. Posakony, L. B. McAllister, R. J. Britten, and E. H. Davidson, "Repetitive Sequences of the Sea Urchin Genome. II. Subfamily Structure and Evolutionary Conservation," *Journal of Molecular Biology* 149 (1981): 15–39; Roy J. Britten, "Rates of DNA Sequence Evolution Differ between Taxonomic Groups," *Science* 231 (1986): 1393–98.

80. Wikipedia, "Gene Silencing," http://en.wikipedia.org/wiki/Gene_silencing, accessed June 3, 2007.

81. Wikipedia, "RNA Interference," http://en.wikipedia.org/wiki/RNAi, accessed June 3, 2007.

82. R. C. Lee, R. L. Feinbaum, and V. Ambros, "The C. Elegans Heterochronic Gene Lin-4 Encodes Small RNAs with Antisense Complementarity to Lin-14," *Cell* 75 (1993): 843–54.

83. G. Ruvkun, "Molecular Biology: Glimpses of a Tiny RNA World," *Science* 294(5543) (October 26, 2001): 797–79.

84. A. Fire, S. Xu, M. Montgomery, S. Kostas, S. Driver, and C. Mello, "Potent and Specific Genetic Interference by Double-Stranded RNA in *Caenorhabditis elegans*," *Nature* 391(6669) (1998): 806–11.

85. John S. Mattick, "A New Paradigm for Developmental Biology," *Journal of Experimental Biology* 210 (2007): 1540.

86. Britten, personal interview with the author.

87. Brian Rotman, *Ad Infinitum . . . The Ghost in Turing's Machine: Taking God Out of Mathematics and Putting the Body Back In* (Stanford, Calif.: Stanford University Press, 1993), 10.

88. "The primary function of writing is to facilitate slavery": Claude Lévi-Strauss, *Tristes tropiques* (Paris: Plon, 1955), 344.

89. In French, the feminine word from the Latin *matricula*, register, metaphorically derived from *mater*, mother (Littré).

90. Norbert Wiener and Arturo Rosenblueth, the fathers of teleonomy, appropriately quoted by Richard Lewontin in his review of Lily Kay's *Who Wrote the Book of Life? Science* 291(5507) (2001): 1263–64.

91. c. 1382, from O.Fr. *gobe* "mouthful, lump," from *gober* "gulp, swallow down," probably from Gaul. **gobbo-* (cf. Ir. *gob* "mouth," Gael. *gob* "beak"). *Online Etymology Dictionary*, entry "gob," http://www.etymonline.com/index.php?term=gob, accessed June 4, 2007.

92. Eugene Thacker, *The Global Genome: Biotechnology, Politics, and Culture* (Cambridge: MIT Press, 2005), 54.

93. Ibid., 60.

94. Nikolas Rose, *The Politics of Life Itself: Biomedicine, Power, and Subjectivity in the Twenty-first Century* (Princeton, N.J.: Princeton University Press, 2007), 9–40.

95. Ibid., 17.

96. Daniel H. Steinberg, "Stein Gives Bioinformatics Ten Years to Live," *O'Reilly Network*, May 5, 2003, http://www.oreillynet.com/pub/a/network/biocon2003/stein.html, accessed June 11, 2007.

97. On this point, see Kaushik Sunder Rajan, on the "Rep-X" case, *Biocapital: The Constitution of Postgenomic Life* (Durham, N.C.: Duke University Press, 2006), 60–67.

98. Michel Houellebecq, *Les particules élémentaires* (Paris: Flammarion, 1998), 23, my translation.

99. William Ray Arney, *Experts in the Age of Systems* (Albuquerque: University of New Mexico Press, 1991). On a different take on this question, see also Peter Sloterdijk, *Critique of Cynical Reason* (Minneapolis: University of Minnesota Press, 1988).

100. Houellebecq, *Les particules élémentaires*, 24, my translation.

101. Britten, personal interview with the author.

102. Gary Stix, "Owning the Stuff of Life," *Scientific American* 294(2) (February 2006): 76–83. Stix's article commented on a recently published paper: Kyle L. Jensen and Fiona Murray, "Intellectual Property Landscape of the Human Genome," *Science* 310 (October 14, 2005): 239–40.

103. According to the jargon file, this "phrase stems from a 1991 adaptation of Toaplan's 'Zero Wing' shoot-'em-up arcade game for the Sega Genesis game console. A brief introduction was added to the opening screen, and it has what many consider to be the worst Japanese-to-English translation in video game history . . . In 2001, this amusing mistranslation spread virally through the Internet, bringing with it a slew of JPEGs and a movie of hacked photographs, each showing a street sign, store front, package label, etc. hacked to read 'All your base are belong to us.'"

104. In *Nobel Lectures, Chemistry 1971–1980*, editor-in-charge Tore Frängsmyr, ed. Sture Forsén (Singapore: World Scientific Publishing Co., 1993), accessed online

at http://nobelprize.org/nobel_prizes/chemistry/laureates/1980/presentation-speech.html.

105. A restriction enzyme (or restriction endonuclease) is a protein that "scans" and cuts double-stranded DNA at a specific site. Werner Arber, Daniel Nathans, and Hamilton Smith earned the 1978 Nobel Prize in medicine for their work in the late 1960s leading to the discovery of restriction endonucleases.

106. A ligase is an enzyme that catalyzes the joining of two large molecules by forming a new chemical bond, usually with accompanying hydrolysis of a small chemical group pendant to one of the larger molecules. B. Weiss and C. C. Richardson isolated the first DNA ligase in 1966.

107. Stanley N. Cohen et al., "Construction of Biologically Functional Bacterial Plasmids in Vitro," *Proceedings of the National Academy of Sciences* 70 (1973): 3240–44.

108. See Wikipedia, "DNA Sequencing," for a more comprehensive *exposé* of the various methods presented hereafter: http://en.wikipedia.org/wiki/DNA_sequencing, accessed June 9, 2007.

109. Frederick Sanger, George G. Brownlee, and Bart G. Barrell, "A Two-Dimensional Fractionation Procedure for Radioactive Nucleotides," *Journal of Molecular Biology* 13(2) (1965): 373–98.

110. Frederick Sanger and Alan R. Coulson, "A Rapid Method for Determining Sequences in DNA by the Primed Synthesis with DNA Polymerase," *Journal of Molecular Biology* 94 (1975): 441–48. "It represented a radical departure from earlier methods in that it did not utilize partial hydrolysis" (*Encylopaedia Britannica*, "Frederik Sanger," http://www.britannica.com/eb/article-260493/Frederick-Sanger, accessed June 9, 2007).

111. Frederik Sanger, Steven Nicklen, and Alan R. Coulson, "DNA Sequencing with Chain-Terminating Inhibitors," *Proceedings of the National Academy of Sciences* 74 (1977): 5463–67.

112. Allan M. Maxam and Walter Gilbert, "A New Method for Sequencing DNA," *Proceedings of the National Academy of Sciences* 74 (1977): 560–64. See also Walter Gilbert, "DNA Sequencing and Gene Structure," in *Nobel Lectures, Chemistry 1971–1980*, 408–26.

113. F. Sanger et al., "Nucleotide Sequence of Bacteriophage phi-X174 DNA," *Nature* 265 (1977): 687–95.

114. R. Staden, "Sequence Data Handling by Computer," *Nucleic Acids Research* 4(11) (1977): 4037–61.

115. Alison McCook, "The Automated DNA Sequencer," *The Scientist* 19(16) (August 29, 2005): 15. See also Merrill Goozner, "Patenting Life," *American Prospect* (December 2000), available on Goozner's own Web site, http://www.gooznews.com/articles/archives/000398.html, accessed June 8, 2007.

116. Walter Gilbert, "Toward a Paradigm Shift in Biology," *Nature* 349 (January 10, 1991): 99.

117. Jean Baudrillard, "The Implosion of Meaning in the Media" (1978), in *In the*

Shadow of the Silent Majorities, trans. Paul Foss, John Johnston, and Paul Patton (New York: Semiotext[e], 1983), 101–2.

118. Jean Baudrillard, *Simulacra and Simulation,* trans. Paul Foss, Paul Patton, and Philip Beitchman (New York, Semiotext[e], 1983 [1981]), 56–57.

119. "The new capitalism is based in sectors not of goods or even services but of *media:* of media as screens, as money and as languages . . . as genetic code in biotechnology. Yet these very media have become products, have become things in an age when capitalism has become metaphysical" (Scott Lash, "Capitalism and Metaphysics," *Theory, Culture and Society* 24[5] [2007]: 19). See chapter 4 for a detailed exposé of this notion.

120. Houellebecq, *Les particules élémentaires,* 392, my translation.

121. See Couze Venn, "Rubbish, the Remnant, Etcetera," *Theory, Culture and Society* 23(2–3) (2006): 44–46.

122. Roger Lewin, "Computer Genome Is Full of Junk DNA," *Science* 232 (May 2, 1986): 577–78.

123. Faye Flam, "Hints of a Language," *Science* 266 (November 25, 1994): 1320.

124. Rachel Nowak, *Science* 268 (February 4, 1994): 798.

125. Diane Gershon, *Nature* 365 (May 18, 1995): 262.

126. Diane Gershon, *Nature* 389 (September 25, 1997): 417–18.

127. Lewin, "Computer Genome Is Full of Junk DNA," 577.

128. Ibid.

129. William F. Loomis and Michael E. Gilpin, "Multigene Families and Vestigial Sequences," *Proceedings of the American Academy of Sciences* 83 (April 1986): 2143.

130. Tomoko Ohta, "Simulating Evolution by Gene Duplication," *Genetics* 115 (January 1987): 212.

131. See Mooto Kimura and Tomoko Ohta, "Protein Polymorphism as a Phase in Molecular Evolution," *Nature* 229 (1971): 467–69, and William F. Loomis, "Vestigial DNA?" *Developmental Biology* 30(2) (1973): f3–f4.

132. Susumu Ohno, see chapter 3.

133. Mooto Kimura, "Evolutionary Rate at the Molecular Level," *Nature* 217 (1968): 624–26.

134. In population genetics, *genetic drift* is the statistical (stochastic) effect that results from the influence that chance has on the survival of variants of a gene (alleles). Its effect may cause an allele to become more common or more rare over successive generations. Along with natural selection, mutation, and migration, it is one of the basic mechanisms of evolution. Whereas natural selection is the tendency of beneficial alleles to become more common over time (and detrimental ones less common), genetic drift is the fundamental tendency of any allele to vary randomly in frequency over time owing to statistical variation alone, so long as it does not comprise all or none of the distribution. With his theory of "forbidden" versus "neutral" mutations, Ohno intuited that after gene or genome duplication, the latter was crucial and natural selection thus turned into genetic drift at the molecular level.

135. James F. Crow, "Motoo Kimura: 13 November 1924–13 November 1994," *Biographical Memoirs of Fellows of the Royal Society* 43 (November 1997): 260; quoted in the "Motoo Kimura" entry of Wikipedia: http://en.wikipedia.org/wiki/Motoo_Kimura, accessed June 12, 2007.

136. "Tests of Neutrality and Selection," from the now defunct "MIT History of Recent Science and Technology Project" at the Dibner Institute, perspectives on molecular evolution, on the Web site of the French Society of Chemistry (Société Française de Chimie), http://www.sfc.fr/material/hrst.mit.edu/hrs/evolution/public/testing.html, accessed June 12, 2007.

137. Rosario N. Mantegna et al., "Linguistic Features of Non-coding Sequences," *Physical Review Letters* 73(23): 3169–72.

138. Ibid., 3172.

139. Flam, "Hints of a Language," 1320. Eugene H. Stanley was one of the principal authors of the *Physical Review Letters* paper.

140. Ibid.

141. N. E. Israeloff, M. Kagalenko, and K. Chan, "Can Zipf Distinguish Language from Noise in Non-coding DNA?" *Physical Review Letters* 76(11): 1976.

142. Sebastian Bonhoeffer et al., "No Signs of Hidden Language in Non-coding DNA," *Physical Review Letters* 76(11): 1977.

143. Richard F. Voss, "Comment on 'Linguistic Features' of Non-coding DNA Sequences," *Physical Review Letters* 76(11): 1978.

144. See, for instance, C. A. Chatzidimitriou, R. M. F. Streffer, and D. Larhammar, "Lack of Biological Significance in the 'Linguistic Features' of Non-coding DNA—A Quantitative Analysis," *Nucleic Acids Research* 24(9) (1996): 1676–81; G. S. Attard, A. C. Hurworth, and J. P. Jack, "Language-like Features in DNA: Transposable Element Footprints in the Genome," *Europhysics Letters* 36(5) (1996): 391–96; H. E. Stanley et al., "Scaling Features of Non-coding DNA," *Physica* 273 (1999): 1–18; Ryuji Suzuki, John R. Buck, and Peter L. Tyack, "The Use of Zipf's Law in Animal Communication Analysis," *Animal Behaviour* 69 (2005): F9–F17; B. Cantu and E. Hernandez-Lemus, "Statistical Properties and Linguistic Coherence in Non-coding DNA Sequences," *Revista Mexicana de Fisica* 51(2) (2005): 118–25.

145. H. Eugene Stanley, "Exotic Statistical Physics: Applications to Biology, Medicine, and Economics," *Physica* A 285 (2000): 3.

146. Chatzidimitriou, Streffer, and Larhammar., "Lack of Biological Significance in the 'Linguistic Features' of Non-coding DNA," 1676, quoting C. Martindale and A. K. Konopka, "Oligonucleotide Frequencies in DNA Follow a Yule Distribution," *Computers and Chemistry* 20(1) (1996): 35–38.

147. Sameer Chouwadhary, "Profiles in Quackery. Holy Qur'an and Genetics," http://www.geocities.com/freethoughtmecca/profiles_in_quackery.htm, accessed June 14, 2007.

148. This one is a spoof, which poses as a genuine "discovery." Last time I checked, I found it on the forum of the *Kuro5hin* Web site "technology and culture, from the

trenches," posted by ake111 under the heading "Searching Human Genome Reveals Biblical Text," http://www.kuro5hin.org/story/2005/4/27/03541/2520.

149. Paul Levinson, "The Copyright Notice Case," first published in *Analog* (1996); this quote, from page 42 of the 2002 Fictionwise e-book edition: http://www.fictionwise.com/ebooks/eBook2304.htm.

150. The Fibonacci sequence of integers (F) is the sequence defined by the following formula: $\forall n \in N, F_{n+1} = F_n + F_{n-1}$. Each number in the sequence is thus the sum of his two anterior numbers: 1,1,2,3,5,8,13,21,34,55, and so on. The sequence is named after the Italian who popularized it in the Western world (the Hindu mathematicians knew it since the sixth century), Leonardo of Pisa, also known as Leonardo Fibonacci (1170 or 1180–1250).

151. Jean-Claude Perez, *L'ADN décrypté: La découverte et les preuves du langage caché de l'ADN* (Embourg: Marco Pietteur, 1997), 18, my translation. This book, only available in New Age bookstores, was nevertheless saluted by some "serious" biologists, such as Jean-Marie Pelt, who wrote its preface.

152. Luc Montagnier, "La musique des gènes," *Science et Avenir* (April 1995).

153. See Mario Livio, *The Golden Ratio: The Story of Phi, the World's Most Astonishing Number* (New York: Broadway Books, 2002).

154. Cf. Martin Gardner, "The Cult of the Golden Ratio," *Skeptical Inquirer* 18 (1994): 243–47.

155. Jean-Claude Perez, personal interview with the author, July 2003.

156. The Chargaff Parity Rule #1 shows that on a double-stranded DNA molecule the percentages of base A and G globally equal, respectively, the percentages of bases T and C. The rigorous validation of this rule constituted the basis of Watson–Crick pair arrangement in their discovery of the structure of DNA in 1953. The Chargaff Parity Rule #2 shows that on a single DNA strand, the percentage of base A globally equals the percentage of base T, and the percentage of base C globally equals the percentage of base G (Wikipedia, "Chargaff's rules"). Both rules combined make Perez's result trivial.

157. Rivka Rudner, John D. Karkas, and Erwin Chargaff, "Separation of B. Subtilis DNA into Complementary Strands, II. Direct Analysis," *Proceedings of the National Academy of Sciences* 60 (1968): 915–20. Chargaff and his colleagues called this "the most surprising result of the analyses reported here" (919), "unexpected compositional regularity evidenced by the single strands" (915), and concluded: "It is certainly advisable to postpone a detailed discussion until we know how universal this relationship is" (919).

158. Günter Albrecht-Buehler, "Asymptotically Increasing Compliance of Genomes with Chargaff's Second Parity Rules through Inversions and Inverted Transpositions," *Proceedings of the National Academy of Sciences* 103 (2006): 17828–33.

159. Michael E. Beleza Yamagishi and Alex Itiro Shimabukuro, "Nucleotide Frequencies in Human Genome and Fibonacci Numbers" (2006), Cornell University Library Open Access Archive of e-prints, http://arxiv.org/abs/q-bio.OT/0611041, accessed

June 18, 2007: "The model relies on two assumptions. First, Chargaff's second parity rule should be valid, and, second, the nucleotide frequencies should approach limit values when the number of bases is sufficiently large. Under these two hypotheses, it is possible to predict the human nucleotide frequencies with accuracy. It is noteworthy, that the predicted values are solutions of an optimization problem, which is commonplace in many of nature's phenomena."

160. Personal interview with the author, 2003.

3. Multimedium, or Life as an Interface Problem

1. Slightly adapted from Robert Musil, *The Man without Qualities*, chapter 72: "Science smiling into its beard, or first full-dress encounter with evil" (1921–42) (London: Secker & Warburg, 1953, 1954, 1960). Musil's man without quality is the grandfather of Homo nexus, this subject without affect (see chapter 5).

2. Terrence and Dennis McKenna, *The Invisible Landscape: Mind Hallucinogens and the I Ching*, rev. ed. (New York: HarperCollins, 1993 [1975]).

3. Ibid., 80–81.

4. Jeremy Narby, *The Cosmic Serpent: DNA and the Origins of Knowledge* (New York: Tarcher/Putnam, 1998), 109.

5. Ibid., 117.

6. Ibid., 30–31.

7. Ibid., 126–27.

8. Ibid., 205n35.

9. Lynne McTaggart, *The Quest for the Secret Force of the Universe* (New York: HarperCollins, 2001).

10. Ibid., xvii. "In physics, the zero-point energy is the lowest possible energy that a quantum mechanical physical system may possess and is the energy of the ground state of the system. The concept of zero-point energy was proposed by Albert Einstein and Otto Stern in 1913, which they originally called 'residual energy' or *Nullpunktsenergie*. All quantum mechanical systems have a zero-point energy . . . Because zero-point energy is the lowest possible energy a system can have, this energy cannot be removed from the system. A related term is zero-point field, which is the lowest energy state of a field, i.e. its ground state, which is non zero" (Wikipedia, "Zero-point energy"). In the linked "Zero-point field" entry, however, Wikipedia adds: "In the area of pseudoscience and popular culture, the zero-point field is an imagined field, vaguely patterned on the concept of the vacuum state in physics, often with appeals to the Casimir effect, that is claimed to have various magical properties . . . In recent years, a number of new age books have begun to appear purporting the view that the zero-point field of physics is the 'secret force' of the universe being used to explain such phenomena as intention, remote viewing, paranormal ability, etc. Books that promote this view include: Lynne McTaggart's *The Field*."

11. McTaggart, *The Field*, 44.

12. Personal interview with the author, Neuss, Germany, June 6, 2003.

13. Tom Bearden, *Excalibur Briefing*, "Biophotons and Virtual EM Field of a Bio-organism," http://www.cheniere.org/books/excalibur/biophotons.htm, accessed June 24, 2007. Note in passing that the McKenna brothers and Narby characterized maninkari as ultraviolet images appearing during the shamanic trance.

14. F. David Peat, "Life and Light," *Noetic Sciences Review* 19 (Autumn 1991): 24, http://www.noetic.org/publications/review/issue19/main.cfm?page=r19_Peat.html, accessed June 25, 2007.

15. Willis Harman, "What Are Noetic Sciences?" *Noetic Sciences Review* 47 (Winter 1998): 32, http://www.noetic.org/publications/review/issue47/r47_Harman.htm, accessed June 25, 2007.

16. "About IONS—Vision and Mission," http://www.noetic.org/about/vision.cfm, accessed June 25, 2007.

17. Rupert Sheldrake, *A New Science of Life: The Hypothesis of Morphic Resonance* (London: Blong and Briggs, 1981). I will refer here to the third edition by Park Street Press (Rochester, Vt., 1995).

18. "A Book for Burning?" *Nature* 293 (September 24, 1981): 246.

19. *Nature* 293 (September 24, 1981): 506, 594, 696, and *Nature* 294 (November 12, 1981): 106.

20. Quoted in Wikipedia, "Rupert Sheldrake," http://en.wikipedia.org/wiki/Rupert_Sheldrake, accessed June 25, 2007.

21. For an account of how Popp, at first a "wiz kid" with no less a bright future in academia, did not get tenure in his first position at the University of Marburg but was rather treated like "a criminal without a fair trial," see McTaggart, *The Field*, 52.

22. Sheldrake, *A New Science of Life*, 22.

23. Rupert Sheldrake, *The Presence of the Past: Morphic Resonance and the Habits of Nature*, 2d ed. (Rochester, Vt.: Park Street Press, 1995), 83 and 87.

24. Personal interview with the author, Montreal, August 2003.

25. It was the case of Alan Sokal (the return of the self-righteous righter) in his case studies of pseudoscientific theories in nursing, "Pseudoscience and Postmodernism: Antagonists or Fellow Travelers?" in *Archeological Fantasies*, ed. Garrett Fagan (2004), http://physics.nyu.edu/faculty/sokal/pseudoscience_rev.pdf, accessed on Sokal's Web site on June 25, 2007.

26. On this issue, see *Beyond Chance and Necessity: A Critical Inquiry into Professor Jacques Monod's Chance and Necessity*, ed. John Lewis (Atlantic Highlands, N.J.: Humanities Press, 1974), and especially C. H. Waddington's contribution, titled "How Much Is Evolution Affected by Chance and Necessity?" 89–102.

27. Conrad Hal Waddington, "The Epigenotype," *Endeavor* 1 (1942): 18–20.

28. Joshua Lederberg, "The Meaning of Epigenetics," *The Scientist* 15(18) (2001): 6.

29. Wikipedia, "Chreode."

30. Sheldrake, *A New Kind of Science*, 12–14.

31. Sheldrake, *The Presence of the Past*, 95.

32. Ibid., 112.

33. Ibid., 113.

34. Sheldrake, *A New Kind of Science,* 52.

35. Personal interview with the author, Montreal, August 2003.

36. Rupert Sheldrake and David Bohm, "Morphogenetic Fields and the Implicate Order," *Re-Vision* 5(2) (1982): 41–48. I will quote here from the large excerpts of this dialogue reprinted in Sheldrake, *A New Science of Life,* 235–47.

37. Ibid., 240.

38. Ibid., 242; editor's emphasis added.

39. The interested reader would profit from a deeper look at the following references that helped me be sure enough to make my claim: David Bohm, *Wholeness and the Implicate Order* (London: Routledge, 1980), especially 186–96 of the 1983 Ark edition; Nick Herbert, *Quantum Reality: Beyond the New Physics* (New York: Anchor Books, 1985), for a general and straightforward presentation of quantum theory interpretations and Bohm's position vis-à-vis realism, neorealism, and Heisenberg's "duplex world of potentials and actualities" (chapter 2, "Physicists Losing Their Grip," 16–29); Robert Nadeau and Menas Kafatos, *The Non-Local Universe: The New Physics and Matters of Mind* (Oxford: Oxford University Press, 1999), especially chapter 9, "Mind Matters: Metaphysics in Quantum Physics" (177–93), for the treatment of the three principal ontologies proposed as alternatives to the reigning Copenhagen interpretation, including de Broglie and Bohm's pilot-wave ontology; and finally, JohnJoe McFadden, *Quantum Evolution: The New Science of Life* (New York: Norton, 2000), for a thorough and clear explanation of the possible relations between molecular biology and quantum physics, and especially his short but incisive description of Bohm's proposition, "The Fairies" (204–6): "Until fairly recently, a belief in hidden variables [Bohm's proposition] was considered the physicist's equivalent of believing in fairies . . . it is better to believe that fairies are the cause of strange goings on at the bottom of your garden than to deny the existence of your garden."

40. McTaggart, *The Field,* 94, 136, and 174.

41. Ibid., 47.

42. Personal interview with the author, Montreal, August 2003.

43. Sheldrake, *A New Science of Life,* 122.

44. Personal interview with the author, Montreal, August 2003.

45. Christa Muths, "Let There Be Light: Why Color Therapy Is Effective," in *Counterpoint: The Personal Development Article Library,* http://www.trans4mind.com/counter point/muths1.shtml, accessed June 29, 2007.

46. See for yourself the thousands of pages that Matti Pitkänen has put on his Web page: http://www.helsinki.fi/~matpitka/.

47. Colm A. Kelleher, "Retrotransposons as Engines of Human Bodily Transformation," *Journal of Scientific Exploration* 13(1) (Spring 1999): 924.

48. Thierry Bardini, *Bootstrapping: Douglas Engelbart, Coevolution and the Origins of Personal Computing* (Stanford, Calif.: Stanford University Press, 2000), 202.

49. Barbara McClintock, "Maize Genetics," *Carnegie Institution of Washington Yearbook* 45 (1946): 180.

50. Evelyn Fox Keller, *A Feeling for the Organism: The Life and Work of Barbara McClintock* (New York: W. H. Freeman, 1983), 178.

51. Following Thorstein Veblen's inspiration. See Erin Wais, "'Trained Incapacity: Thorstein Veblen and Kenneth Burke," *K. B. Journal* 2(1) (fall 2005), http://kbjournal.org/node/103/72, accessed June 30, 2007.

52. Keller, *A Feeling for the Organism*, chapter 8, "Transposition Rediscovered," 171–95.

53. Barbara McClintock, "The Significance of Responses of the Genome to Challenge," in *Nobel Lectures: Physiology or Medicine 1981–1990*, editor-in-charge Tore Frängsmyr, ed. Jan Lindsten (Singapore: World Scientific Publishing Co., 1993), 198, emphasis added.

54. Margaret Kidwell, "Transposable Elements," in *The Evolution of the Genome*, ed. T. Ryan Gregory (Amsterdam: Elsevier, 2005), 189.

55. Frederic Bushman, *Lateral DNA Transfer: Mechanisms and Consequences* (New York: Cold Spring Harbor Laboratory Press, 2002), 431.

56. For example, T. Ryan Gregory, "Genome Size Evolution in Animals," in Gregory, *The Evolution of the Genome*, 28.

57. Kidwell, "Transposable Elements," 205.

58. James A. Shapiro, "A 21st Century View of Evolution: Genome System Architecture, Repetitive DNA, and Natural Genetic Engineering," *Gene* 345 (2005): 93.

59. James A. Shapiro, "A Third Way," *Boston Review* (February/March 1997), http://www.bostonreview.net/BR22.1/shapiro.html, accessed June 30, 2007.

60. Keller, *A Feeling for the Organism*, 202.

61. Ibid., 205.

4. Close Encounters of the Fourth Kind

1. Annalee Newitz, "Home-Built Honeybees—Hold the Venom," *Wired* online Web site, http://www.wired.com/wired/archive/10.06/start.html?pg=18, accessed April 28, 2008.

2. Graeme Smith, "Baby Said to Be Clone Will Vanish, Raelian Announces," *Globe and Mail*, January 24, 2003.

3. I first thought it was a hoax, but the product exists; it is available at the online store of the Discovery channel: http://shopping.discovery.com/stores/servlet/ProductDisplay?catalogId=10000&storeId=10000&productId=53965&langId=-1&search=Y&searchKey=-471800637.

4. Alison McCook, "UN Bans Reproductive Cloning: Non-legally Binding Agreement Doesn't Explicitly Address Therapeutic Cloning," *The Scientist*, February 21, 2005.

5. "Although he wasn't up for election, you could say that CALIFORNIA Gov. Arnold Schwarzenegger was among the big winners on election night . . . at least in some areas. Voters endorsed 11 of the 15 ballot measures on which Schwarzenegger

took a stand, including controversial measures on embryonic stem cell research and revising the state's three-strikes law ("2004 Elections," *State Net Capitol Journal* 12[44] [November 8, 2004], http://www.legislate.com/capj/capj.cgi?issue=20041108).

6. "Post-scriptum sur les sociétés de contrôle," *L'autre journal* 1 (1990), reprinted in *Pourparlers* (Paris: Minuit, 1990), 240–47, and first translated in *October* 59(3–7) (1992) as "Postscript on the Societies of Control." I use the French original in *Pourparlers* (with page indications) and update the translation provided in *October.*

7. Ibid., 244. The *October* translation is "capable of generating them and using them."

8. Gilles Deleuze and Félix Guattari, *A Thousand Plateaus: Capitalism and Schizophrenia,* trans. Brian Massumi (Minneapolis: University of Minnesota Press, 1987), 456–57.

9. Walt Whitman Rostow, *The Stages of Economic Growth: A Non-Communist Manifesto* (Cambridge: Cambridge University Press, 1960). Rostow describes five stages of growth: (1) traditional societies, (2) the societies in process of transition when the preconditions of takeoff are developed, (3) the takeoff, (4) the drive to maturity, and (5) the age of mass consumption. This model has been rightly criticized for its linearity and its Occidental ethnocentrism (among others, by Samir Amin and André Gunder Frank). It has nevertheless one quality worth noticing here: it differentially dates its stages according to national development.

10. Deleuze, "Postscript on the Societies of Control," 246.

11. Gilles Deleuze, "Contrôle et devenir," in *Pourparlers,* 229–39, my translation.

12. Gilbert Simondon, *Du mode d'existence des objets techniques* (Paris: Aubier, 1989 [1958]), 159. Jean-Hughes Barthélémy provides the full sense of this concept in his section titled "first difficulty: from 'allagmatics' to the 'theory of the phases of being'" in his definitive *Penser la connaissance et la technique après Simondon* (Paris: L'Harmattan, 2005), 99–111.

13. "We [Guattari and Deleuze] do not believe in a political philosophy that would not be centered on the analysis of capitalism and its developments" (Deleuze, "Contrôle et devenir," 232).

14. Deleuze, "Postscript on the Societies of Control," 245.

15. Deleuze, "Contrôle et devenir," 237.

16. See William Burroughs, "DEPOSITION: Testimony Concerning a Sickness," in *Naked Lunch* (New York: Grove Press, 1990 [1959]), xxxv–xlv, and especially this other "post-script . . . Wouldn't you?": "Do I hear muttering about a personal razor and some bush league short con artist who is known to have invented The Bill? Wouldn't You? The razor belonged to a man named Occam and he was not a scar collector. Ludwig Wittgenstein *Tractatus Logico-Philosophicus:* 'If a proposition is NOT NECESSARY is it MEANINGLESS and approaching MEANING ZERO.' 'And what is More UNNECESSARY than junk if You Don't Need it?' *Answer:* 'Junkies, if you are not ON JUNK.'" (xliii). I'll make mine his slogan for the depiction of genetic capitalism: "Room for one more inside, Sir" (see chapter 6).

17. Ibid., xxxvi.

18. Deleuze, "Postscript on the Societies of Control," 246.

19. "A notion of ethics has to be seen not as an incidental element of Deleuze's project but one of its most fundamental and essential elements. Deleuze is, in fact, compelled by the very adventure of thought to think ethically and even to think an ethics of matter itself" (Keith Ansell Pearson, *Germinal Life: The Difference and Repetition of Deleuze* [London: Routledge, 1999], 11).

20. Slavoj Žižek, "Against Hyphen Ethics," in *Organs without Bodies: On Deleuze and Consequences* (New York, Routledge, 2004), 123–33.

21. Ibid., 126.

22. Ibid., 133.

23. Gilles Deleuze, *Foucault* (Minneapolis: University of Minnesota Press, 1988), 132.

24. Deleuze does not say where in Rimbaud: it is in his letters, the "Lettres dites du voyant": "Donc le poète est vraiment voleur de feu. Il est chargé de l'humanité, des *animaux* même" (Lettre à Georges Izambard, May 1871). Overman is thus first a poet, a Promethean poet ready for a season in hell.

25. Ansell Pearson, *Germinal Life,* 222.

26. See Friedrich Nietzsche, *The Ass Festival.* Loaded as in "intoxicated," drunk. Some insights from Rimbaud again: "Here comes the time of the assassins" ("Matinée d'ivresse"), "Here no hope / no *orietur* / science with patience / torture for sure" ("L'éternité," in *Festivals of Patience,* my translation).

In *Difference and Repetition,* Deleuze notes: "Thus Zarathustra's Ass says yes; but to him, to affirm means to bear, to assume or to shoulder a burden oneself. He bears everything: the burdens with which he is laden (divine values), those which he assumes himself (human values), and the weight of his tired muscles when he no longer has anything to bear (the absence of values). This Ass and the dialectical ox leave a moral aftertaste. They have a terrifying taste for responsibility, as though it were necessary to pass through the misfortunes of rifts and division in order to be able to say yes" (trans. Paul Patton [New York: Columbia University Press, 1994], 53).

27. Gilles Deleuze and Félix Guattari, *Anti-Oedipus: Capitalism and Schizophrenia,* trans. Robert Hurley, Mark Seem, and Helen R. Lane (Minneapolis: University of Minnesota Press, 1983), 4.

28. Rimbaud again, in "Soleil et chair" (Sun and flesh), originally titled "*Credo in unam*" (*Arthur Rimbaud, Collected Poems,* trans. Olivier Bernard, 1962): "If only the times which have come and gone might come again! /—For Man is finished! Man has played all the parts! / In the broad daylight, wearied with breaking idols / He will revive, free of all his gods, / And, since he is of heaven, he will scan the heavens!"

29. I borrow the expression "abstract sex" from Luciana Parisi's eponymous book.

30. "Yeshua said, 'When you see one not born of woman, fall on your faces and worship. That is your Father'" (*Gospel of Thomas,* in *The Gnostic Bible,* ed. Willis Barnstone and Marvin Meyer [Boston: Shambhala, 2003], 49).

31. Žižek, "Against Hyphen Ethics," 124.

32. "Yeshua said, 'Blessings on the lion if a human eats it, making the lion human. Foul is the human if a lion eats it, making the lion human'" (*Gospel of Thomas*, 46).

33. Alain de Libera, *La querelle des universaux: De Platon à la fin du Moyen Âge* (Paris: Seuil, 1996), 335–36.

34. Deleuze, *Difference and Repetition*, 35.

35. Ibid., 39–40.

36. Ibid., 40–41.

37. Ibid., 41.

38. Laurence Paul Hemming, "Nihilism: Heidegger on the Grounds of Redemption," in *Radical Orthodoxy: A New Theology*, ed. John Milbank, Catherine Pickstock, and Graham Ward (New York: Routledge, 1999), 98. And Pierre Klossowski adds, alas, that the subject (starting with this subject called Nietzsche) is hitherto dispelled. Overman is a necessary consequence of the impossibility of the subject to take on such a load: "It is when he announces that *all the gods are dead* that Zarathustra demands that from this point the *overman live*, that is, the humanity which would know how to overcome itself. How does it overcome itself? In desiring again that all things that were already reproduced themselves, and this as its proper/own activity: this action is defined as will to creation and Zarathustra declares that *if there were gods, what would there be to create?* But what is it that brings man to create if not precisely the law of the eternal return to which he decides to adhere? What does he adhere to if it is not precisely a life that he has *forgotten*, but that the revelation of the eternal return as law incites him to want again? And what does he want again if it is not precisely what just now he did not think of wanting: is this saying that the absence of gods incites him to create new gods? Or rather does he want to prevent the return of the ages when he adored the gods? By re-willing the gods, now that he wants the man's passage to a superior life? But how else, given this, would this life be superior, if not by tending towards what it was already? How, in other words, if not by tending towards this state in which it thought nothing of creating, but rather of adoring the gods? And thus it appears that the doctrine of the eternal return is once again conceived as a *simulacrum of a doctrine* whose very parodic character accounts for *hilarity* as a self-sufficient attribute of existence when the laughter bursts at the foundation of truth in its entirety, whether it is that the truth explodes in the laughter of the gods, or whether the gods themselves die of mad laughter: 'When a god wanted to be the only God, all the other gods were gripped by mad laughter to the point of *dying* of laughter. For what is the divine if not the fact that there be several gods and not God alone?'" Pierre Klossowski might give the final clue in his wonderful "Nietzsche, Polytheism and Parody," in *Un si funeste désir* (Paris: Gallimard, 1963), revised translation adapted from at http://lists.village.virginia.edu/cgi-bin/spoons/archive1.pl?list=deleuze-guattari.archive/papers/kloss.polytheism.

Remember that "the eternal return" was first an experience and not a concept for Nietzsche (an experience where he lost his mind and re-created it anew).

39. And with Paolo Virno's injunction to think about common nature and singularity at the light of Duns Scotus and Gilbert Simondon's works: "Les anges et le général intellect: L'individuation chez Duns Scot et Gilbert Simondon," *Multitudes* 18 (2004): 33–45.

40. Gerard Sondag, "Introduction," in Duns Scot, *Le principe d'individuation* (Paris: Vrin, 1992), 35–36.

41. Jorge J. E. Gracia, *Individuality: An Essay on the Foundations of Metaphysics* (Albany: State University of New York Press, 1988).

42. Gilbert Simondon, *L'individuation à la lumière des notions de forme et d'information* (Grenoble: Millon, 2005).

43. "Moi, Antonin Artaud, je suis mon fils, mon père, ma mère, et moi; niveleur du périple imbécile où s'enferre l'engendrement" (Antonin Artaud, "Ci-gît," in *Œuvres* (Paris: Gallimard, 2004 [1946]), 1152.

44. Brian Massumi, *A User's Guide to Capitalism and Schizophrenia: Deviations from Deleuze and Guattari* (Cambridge: MIT Press, 1992), 83, emphasis added. See Deleuze and Guattari, "The Molecular Unconscious," in *Anti-Oedipus*, 283–96, for the original claims, and especially their retake of Lipot Szondi's "genic unconscious."

45. See John Marks, "Molecular Biology in the Work of Deleuze and Guattari," *Paragraph* 29(2) (2006): 83, for an elaboration of this point: "Jacob's interpretation of molecular biology also converges with Deleuze and Guattari's more general conviction . . . that extensive organic forms in fact a single organic stratum, a continuum."

46. Deleuze and Guattari, *Anti-Oedipus*, 288.

47. Deleuze and Guattari attribute this insight to Marx himself: "But Marx says something even more mysterious: that the true difference is not the difference between the two sexes, but the difference between the human sex and the 'nonhuman' sex" (ibid., 294).

5. Lysis and Replication

1. This expression refers to the status of a Web site that has been completely removed as a result of net abuses by the Web site operators.

2. The following biographical material is a synthesis from various sources, including H. L. Drake, *A. E. van Vogt: Science Fantasy's Icon Author,* Roger Russell's A. E. van Vogt page (http://www.roger-russell.com/sffun/nulla.htm); James Gunn, *Introduction to the World of Null-A* (Norwalk, Conn.: Easton Press, 1995), whose introduction is available at http://vanvogt.www4.mmedia.is/intronul.htm; and Alexei Panshin, *Man Beyond Man: The Early Stories of A. E. van Vogt* (http://www.enter.net/~torve/articles/vanvogt/vanvogt1.html).

3. This story provided the inspiration for Ridley Scott's 1979 movie *Alien.* First unacknowledged, the influence got recognized after van Vogt filed a lawsuit claiming plagiarism. The lawsuit was settled out of court, and van Vogt got both an undisclosed sum of money and a presence in the credits of the movie.

4. On"Nexialism," see http://www.nexial.org/nexialism.htm.

5. A. E. van Vogt, *The Voyage of the Space Beagle* (New York: Macmillan Collier, 1992), 60.

6. Alfred Korzybski, *Science and Sanity: An Introduction to Non-Aristotelian Systems and General Semantics*, 4th ed. (Lakeville, Conn.: International Non-Aristotelian Library Publishing Company, 1958 [1933]), i.

7. Alfred North Whitehead, *Process and Reality*, corrected ed. (New York: Free Press, 1978 [1929]), 20.

8. Ibid., 18.

9. Ibid., 20.

10. See John W. Lango, "Whitehead's Category of Nexus of Actual Entities," *Process Studies* 29(1) (2000): 16–42, http://www.religiononline.org.

11. Whitehead, *Process and Reality*, 22.

12. This seems very close to William James's program for "radical empiricism," which states that "relations that connect experiences must themselves be experienced relations, and any kind of relation experienced must be accounted as 'real' as anything else in the system" (William James, *Essays on Radical Empiricism* [New York: Dover Publications, 2003 (1912)], 23, emphasis in the original).

13. Whitehead, *Process and Reality*, 73.

14. Lango, "Whitehead's Category of Nexus of Actual Entities."

15. Gilles Deleuze, Course on Leibniz, Vincennes, March 17, 1987, available online at http://www.webdeleuze.com/php/texte.php?cle=142&groupe=Leibniz&langue=1, accessed May 3, 2008.

16. George Long, "Nexum," in *Dictionary of Greek and Roman Antiquities*, ed. William Smith (London: John Murray, 1875), 795–98, http://penelope.uchicago.edu/Thayer/E/Roman/Texts/secondary/SMIGRA*/Nexum.html.

17. "'*Mancipium*' or mancipation was a formal public ceremony required for recognition of conveyance in 'title' of legal ownership to a thing (*mancipatio*—taking in hand). The ceremony included striking a scale with a copper ingot as a token of sale. Without this ancient ritual, no exchange had the sanction or protection of the law" (George Long, "Mancipium," in *Dictionary of Greek and Roman Antiquities*, 727–28), http://penelope.uchicago.edu/Thayer/E/Roman/Texts/secondary/SMIGRA*/Mancipium.html.

18. My translation with the help of the French translation from J. Gaudemet, *Droit privé romain*, 2d ed. (Paris: Montchrestien, 2000), 385–96n136. Bronze here refers to the "copper ingots" of the description of the mancipio ritual: bronze (a copper-based alloy) ingots were the ritual intermediary used before there was money (coins).

19. http://www.m-w.com/cgi-in/dictionary?book=Dictionary&va=nexus.

20. http://www.bartleby.com/61/7/N0090700.html.

21. http://www.bartleby.com/61/roots/IE338.html.

22. See Sadie Plant, "Mobile Knitting," in *Information Is Alive* (Rotterdam: V2_/NAi Publishers, 2003), 26–37.

23. Richard Doyle, "LSDNA: Rhetoric, Consciousness Expansion, and the Emergence of Biotechnology," *Philosophy and Rhetoric* 35(2) (2002): 153–74.

24. Gilles Deleuze and Félix Guattari, *A Thousand Plateaus: Capitalism and Schizophrenia*, trans. Brian Massumi (Minneapolis: University of Minnesota Press, 1987), 448–49.

25. Georges Dumézil, *Mitra-Varuna: Essai sur deux représentations indo-européennes de la souveraineté*, 3d ed. (Paris: Gallimard, 1948), 132. The first edition of this book appeared in 1940, and it was based on some conferences given in 1938–39.

26. Deleuze and Guattari, *A Thousand Plateaus*, 428–29.

27. Ibid.

28. Ibid., 449–453.

29. Ibid., 460.

30. Ibid.

31. Karl Marx, *Economic and Philosophic Manuscripts of 1844*, quoted in ibid., 129.

32. Deleuze and Guattari, *A Thousand Plateaus*, 460.

33. Marcel Mauss, *Sociologie et anthropologie*, 6th ed. (Paris: Presses Universitaires de France, 1995), 232.

34. Dumézil, *Mitra-Varuna*, 118.

35. Ibid.

36. Ibid., 121.

37. "A man advanced in years rushed into the forum with the tokens of his utter misery upon him . . . Notwithstanding his wretched appearance however, he was recognised, and people said that he had been a centurion, and, compassionating him, recounted other distinctions that he had gained in war: he himself exhibited scars on his breast in front, which bore witness to honourable battles in several places. When they repeatedly inquired the reason of his plight, and wretched appearance, a crowd having now gathered round him almost like a regular assembly, he said, that, while serving in the Sabine war, because he had not only been deprived of the produce of his land in consequence of the depredations of the enemy, but his residence had also been burned down, all his effects pillaged, his cattle driven off, and a tax imposed on him at a time when it pressed most hardly upon him, he had got into debt: that this debt, increased by exorbitant interest, had stripped him first of his father's and grandfather's farm, then of all his other property; lastly that, like a wasting sickness, it had reached his person: that he had been dragged by his creditor, not into servitude, but into a house of correction and a place of torture. He then showed his back disfigured with the marks of recent scourging. At this sight and these words a great uproar arose" (Titus Livius, *Roman History*, Book II, trans. J. H. Freese, A. J. Church, and W. J. Brodribb [1904], http://ancienthistory.about.com/library/bl/bl_text_livy_2.htm, accessed December 11, 2006).

38. Jean Baudrillard, "The Mirror of Production," in *Utopia Deferred: Writings for Utopie (1967–1978)*, trans. Stuart Kendall (New York: Semiotext[e], 2006), 130.

39. Gilles Deleuze and Félix Guattari, *Anti-Oedipus: Capitalism and Schizophrenia*, trans. Robert Hurley, Mark Seem, and Helen R. Lane (Minneapolis: University of Minnesota Press, 1983), 289–94.

40. Jean Baudrillard, "Immortality," in *The Illusion of the End* (Stanford, Calif.: Stanford University Press, 1994), 95.

41. Ibid., 97.

42. Pierre Klossowski, *La monnaie vivante* (Paris: Rivages poche, "Petite bibliothèque," 1997 [1970]), 75–77, my translation.

43. Friedrich Nietzsche, *Twilight of the Idols* (Harmondsworth: Penguin, 1968), 41.

44. BBC News: http://news.bbc.co.uk/1/hi/health/3545684.stm, August 8, 2004.

45. "The organic composition of man refers by no means only to his specialized technical faculties, but . . . equally to their opposite, the moments of naturalness which once themselves sprung from the social dialectic and are now succumbing to it. Even what differs from technology in man is now being incorporated into it as a kind of lubrication. Psychological differentiation, originally the outcome both of the division of labour that dissects man according to sectors of the production process and of freedom, is finally itself entering the service of production" (Theodor Adorno, "Novissimum organum," in *Minima moralia,* trans. E. F. N. Jephcott [London: Verso, 1974 (1951)], 229–30).

46. Bernard Stiegler, *Mécréance et discrédit II: Les sociétés incontrôlables d'individus désaffectés* (Paris: Galilée, 2006), 14–15, my translation.

47. See especially Giorgio Agamben's chapter 9, "In Praise of Profanation," in *Profanations* (New York: Zone Books, 2007), 73–92.

48. Walter Benjamin,"Capitalism as Religion," in *Selected Writings,* vol. 1, *1913–1926,* ed. Marcus Bullock and Michael W. Jennings (Cambridge: Belknap Press of Harvard University Press, 1996), 288–91. Subsequent references are given in the text.

49. Agamben, "In Praise of Profanation," 81.

50. Ibid., 84.

51. Günther Anders, *L'obsolescence de l'homme: Sur l'âme à l'époque de la deuxième révolution industrielle* (Paris: Encyclopédie des Nuisances, 2002 [1956]).

52. Benjamin, "Capitalism as Religion," 289.

53. Agamben, "In Praise of Profanation," 92.

54. Marcel Detienne and Jean-Pierre Vernant, *Les ruses de l'intelligence: La Métis des Grecs* (Paris: Flammarion, 1974), 93–96.

55. Plato, *Protagoras* 321d.

56. Aeschylus, *Prometheus Bound,* 463–64, http://etext.library.adelaide.edu.au/mirror/classics.mit.edu/Aeschylus/prometheus.html, accessed December 14, 2006.

57. Ibid.

58. Plato, *Protagoras* 322b.

59. Ibid., 322c.

60. Aeschylus, *Prometheus Bound,* 967.

61. Hesiod, *Works and Days,* trans. Bruce MacLennan (1995), 45–103, http://www.stoa.org/diotima/anthology/hes_pandora.shtml, accessed December 15, 2006.

62. As in *The Da Vinci Code:* she is the Holy Grail.

63. Jacques Lacan, *Séminaire X: L'angoisse* (Paris: Seuil, 2004), 187.

64. Ibid., 195.

65. See Irina Aristarkhova, "Ectogenesis and Mother as Machine," *Body and Society* 11(3) (2005): 43–59.

66. Jean-Pierre Vernant, "À la table des hommes: Mythe de fondation du sacrifice chez Hésiode," in Marcel Detienne and Jean-Pierre Vernant, *La cuisine du sacrifice en pays grec* (Paris: Gallimard, 1979), 131–32, my translation. One will consult with great interest the use of this source and its interpretation of the Promethean myth in Bernard Stiegler's first volume of *Technics and Time: The fault of Epimetheus* (Stanford, Calif.: Stanford University Press, 1998), 188–203. And especially this quote translated from Vernant: "if, by placing Elpis in the jar (of Pandora) in company with all ills, Hesiod completely assimilated it, making it into the anxious expectation of ill in order to avoid ambiguity, ought he not to have named it Phobos in preference to Elpis?" (Stiegler, *Technics and Time*, 197). Phobos and his twin brother Deimos, sons of Ares, personifications of Fear and Panic, are the closest ills to Anxiety. They definitely fled among men when the box was opened!

67. Peter Sloterdijk, *La compétition des bonnes nouvelles: Nietzsche évangéliste* (Paris: Mille et Une Nuits, 2002), 84, my translation.

68. Tom Peters, "The Brand Called You," *Fast Company* 10 (August 1997): 83. I do not resist giving the header of this article: "Big companies understand the importance of brands. Today, in the Age of the Individual, you have to be your own brand. Here's what it takes to be the CEO of Me Inc."

69. Lacan, *Séminaire X*, 65.

70. Patrick Barry, "Stem Cells Turned into Sperm," *New Scientist* (July 15, 2006), http://www.newscientist.com/channel/sex/stem-cells/mg19125604.200-stem-cells-turned-into-sperm.html, accessed December 17, 2006.

71. "Spermatozoïdes synthétiques: La science rejoint la fiction," *La Presse*, cahier actuel, July 16, 2006, 7.

72. Mark Henderson, "Fathers Are Out of the Picture as Lesbians Get IVF," *Times*, July 13, 2006, http://www.timesonline.co.uk/article/0,,17129-2267632,00.html, accessed December 17, 2006.

73. Samuel Butler, "The Book of the Machines," chapters 23–25 of *Erewhon* (London: Penguin Classics, 1985 [1872]), 210–11.

74. Deleuze and Guattari, *Anti-Oedipus*, 285.

75. Ibid.

76. John G. Ballard, "Project of a Glossary of the Twentieth Century" (in *Zone 6: Incorporations*, edited by Jonathan Crary and Sanford Kwinter, New York: Zone, 1992), 271.

77. Linda Brigham, "Are We Posthuman Yet? Review of N. Katherine Hayles's *How We Became Posthuman*," *AltX*, http://www.altx.com/ebr/reviews/rev9/r9bri.htm, accessed December 15, 2006.

78. Katherine N. Hayles, *How We Became Post-Human* (Chicago: University of Chicago Press, 1999), 33.

79. Ibid., 43.

80. Ibid., 32.

81. Deleuze and Guattari, *Anti-Oedipus*, 295.

82. Aldous Huxley, *Brave New World* (1932; London: Penguin Modern Classics, 1976), 30.

83. Michel Houellebecq, *Les particules élémentaires* (Paris: Flammarion, 1998), 385, my translation.

84. Ibid., 199–200.

85. Michel Houellebecq, *La possibilité d'une île* (Paris: Fayard, 2005), 392, my translation.

86. Albert Camus, *The Myth of Sisyphus* (New York: Vintage Books edition, 1991), 3.

87. Ibid., 123.

6. Presence of Junk

1. Emmanuel Carrère, *I Am Alive and You Are Dead: A Journey into the Mind of Philip K. Dick* (London: Picador, 2005).

2. "Outline in Abstract Form of a New Model of Reality Updating Historic Models (in Particular Those of Gnosticism and Christianity," an Exegesis entry dated December 7, 1977, and cosigned by K. W. Jeter, in *The Shifting Realities of PKD: Selected Literary and Philosophical Writings*, ed. Lawrence Sutin, (New York: Vintage Books, 1995), 326–27.

3. From a letter PKD sent to one of his critics, dated December 9, 1967, available on the *Total Dick Head* blog, http://totaldickhead.blogspot.com/2007/06/satanic-bible-eldritch-reviewed.html, accessed August 18, 2007.

4. Katherine Hayles, "Escape and Constraint: Three Fictions Dream of Moving from Energy to Information," in *From Energy to Information: Representations in Science and Technology, Art, and Literature*, ed. Bruce Clarke and Linda Dalrymple Henderson (Stanford, Calif.: Stanford University Press, 2002), 249.

5. Philip K. Dick, quoted in *In Pursuit of Valis: Selections from The Exegesis*, ed. Lawrence Sutin, (Lancaster, Pa.: Underwood-Miller, 1991), 185 of the Vintage edition.

6. *The American Heritage Dictionary of the English Language*, 4th ed. (2000) online at http://www.bartleby.com/61/71/S0757100.html.

7. Philip K. Dick, *The Three Stigmata of Palmer Eldritch* (New York: Vintage, 1991 [1964], 161–62 of the Vintage edition.

8. Hayles, "Escape and Constraint," 249.

9. Dick, *The Three Stigmata of Palmer Eldritch*, 224 of the Vintage edition.

10. Erik Davis, "Philip K. Dick's Divine Interference," online at http://www.techgnosis.com/pkdnet.html.

11. Alexander Star, "The God in the Trash: The Fantastic Life and Oracular Work of Philip K. Dick," *New Republic* 209(23), December 6, 1993, online at http://www.popsubculture.com/pop/bio_project/philip_k_dick.2.html.

12. Philip K. Dick, *Do Androids Dream of Electric Sheep?* (first published by Del Rey in 1968) (New York: Ballantine, 1982), 143.

13. Judith B. Kerman, "Technology and Politics in Blade Runner Dystopia," in *Retrofitting* Blade Runner*: Issues in Ridley Scott's* Blade Runner *and Philip K. Dick's* Do Androids Dream of Electric Sheep?, ed. Judith B. Kerman (Bowling Green, Ohio: Bowling Green State University Popular Press, 1991), 18.

14. Dick, *Do Androids Dream of Electric Sheep?,* 15.

15. Ibid., 13.

16. Cf. Giulina Bruno, "Ramble City: Postmodernism and *Blade Runner,*" *October* 41 (1987) 61–74, who notes: "There is even a character in the film who is nothing but a literalization of this condition" (65).

17. Dick, *Do Androids Dream of Electric Sheep?,* 12.

18. Ibid., 54.

19. Ibid., 57.

20. Ibid., 57–58.

21. This word occurs in other PKD novels, such as in *A Maze of Death* (1970), for instance.

22. Stephen Nottingham, *Screening DNA: Exploring the Cinema-Genetics Interface* (1999), available at http://ourworld.compuserve.com/homepages/Stephen_Notting ham/DNA4.htm.

23. And when Sebastian meets Pris, she is hidden under layers of garbage in front of his apartment building.

24. In a fascinating variation on the themes of both the golem and the fallen angel. Cf. David Desser, "The New Eve: The Influence of Paradise Lost and Frankenstein on Blade Runner," in Kerman, *Retrofitting* Blade Runner, 53–65.

25. See Joseph Francavilla, "The Android as a Doppelgänger," in Kerman, *Retrofitting* Blade Runner, 4–15, for a further elaboration of the replicant (and especially Roy Batty, their leader) as man-made *Übermensch.*

26. Scott Bukatman, "Filming Blade Runner," http://www.bfi.org.uk/bookvid/books/catalogue/sample/text.php?bookid=44.

27. Gary Indiana "Burroughs," http://www.criterionco.com/asp/release.asp?id=220&eid=333§ion=essay&page=6.

28. William S. Burroughs, *Blade Runner: A Movie* (Berkeley: Blue Wind Press, 1989), not paginated.

29. Ibid.

30. Ibid.

31. "One must further recognize and accept the pervasiveness of the viral trope within postmodernism . . . and understand the ontological confusion (and ideological anxiety) which it carries" (Scott Bukatman, *Terminal Identity: The Virtual Subject in Postmodern Science Fiction* [Durham, N.C.: Duke University Press, 1993], 347).

32. For this purpose, I will draw heavily on the vocabulary of the Diffusion of Innovation theory (logistic curve, inflexion points, critical mass, etc.). Ironically enough, the logistic model of the diffusion of innovations was originally borrowed from the field

of epidemiology. See Everett M. Rogers, *Diffusion of Innovations*, 4th ed. (New York: Free Press, 1995).

33. Incidentally, Salvador E. Luria, Max Delbrück and Alfred D. Hershey were awarded the Nobel Prize in physiology or medecine for their work on viruses in 1969. As early as 1955, Salvador Luria had written that "A new view of the nature of viruses is emerging. They used to be thought of solely as foreign intruders—strangers to cells they invade and paratize. But recent findings, including the discovery of host-induced modifications of viruses, emphasize more and more the similarity of viruses to hereditary units such as genes. Indeed, some viruses are being considered as bits of heredity in search of a chromosome" ("50, 100 & 150 Years Ago," *Scientific American* [April 2005]: 18).

34. William S. Burroughs, *The Electronic Revolution* (1970, Expanded Media Editions published by Bresche Publikationen Germany), English version available at http://www.hyperreal.org/wsb/elect-rev.html.

35. Kathy Acker, "Returning to the Source," funeral oration for William Burroughs (*21C* 26 [1998]: 14): "He was the detective. Being the detective, he was the doctor. He searched out the possessors."

36. Ibid.

37. Cf. Anne-Marie Christin, *L'image écrite, ou la déraison graphique* (Paris: Flammarion, 2001 [1995]).

38. Jacques Derrida, *Of Grammatology* (Baltimore: Johns Hopkins University Press, 1977 [1967]), 314.

39. Jacques Derrida with Peter Brunette and David Wills, "The Spatial Arts: An Interview with Jacques Derrida," in *Deconstruction and the Visual Arts: Art, Media Architecture,* ed. Peter Brunette and David Wills (Cambridge: Cambridge University Press, 1994), 12.

40. Richard Dawkins, *The Selfish Gene* (London and New York: Norton, 1989 [1976]), 192.

41. There are instances of "viral infections" documented for the Univac 1108 and the IBM 360/370 ("Pervading Animal" and "Christmas Tree").

42. For more info on Elk Cloner, see http://www.skrenta.com/cloner.

43. Susan Sontag, "Illness as Metaphor," in *Illness as Metaphor and AIDS and Its Metaphors* (London: Picador, 1990 [1978]), 63.

44. See René Girard, *Des choses cachées depuis la fondation du monde* (Paris: Grasset, 1978) and *Le Bouc Émissaire* (Paris: Grasset, 1982).

45. Gabriel de Tarde, *Les lois de l'imitation* (1890), for instance.

46. Lorenzo Miglioli, "Berlusconi Is a Retrovirus: From the Italian Theory-Fiction Novel," in *Digital Delirium*, ed. Arthur and Marilouise Kroker (Montreal: New World Perspectives, 1997), 145.

47. Frederic Cohen, "Computer Viruses—Theory and Experiments," DOD/NBS 7th Conference on Computer Security, originally appearing in IFIP-sec 84, also appearing

in "Computers and Security," V6(1987), 22–35, and other publications in several languages: http://vx.netlux.org/lib/afc01.html.

48. "Innerview," http://www.scu.edu.cn/waim03/scu_cs/teach/adleman.htm.

49. Ibid.

50. Laurie Anderson, "Language Is a Virus," *Home of the Brave* (1986).

51. Prince, "Signs O' the Times," *Signs O' the Times* (1987).

52. Susan Sontag, "AIDS and Its Metaphors," in *Illness as Metaphor and AIDS and Its Metaphors*, 114. And she adds: "I am thinking, of course, of the United States, where people are currently being told that heterosexual transmission is extremely rare, and unlikely—as Africa did not exist."

53. The Pet Shop Boys, "It Couldn't Happen Here," *Actually* (1987).

54. Bruce Springsteen, "Streets of Philadelphia," *Philadelphia Soundtrack* (1993).

55. Burroughs, *The Electronic Revolution.*

56. It seems quite ironic again that the controversy about the HIV "hypothesis" should have exploded right at the time the hypervirus pandemic passed its final inflexion point, around 1993. In 1993, Kary Mullis, in an interview for the *Sunday Times*, said: "If there is evidence that HIV causes AIDS, there should be scientific documents which either singly or collectively demonstrate that fact, at least with a high probability. There is no such document" (November 28, 1993). A year later, again in the *Sunday Times*, Dr. Bernard Forscher, former editor of the U.S. *Proceedings of the National Academy of Sciences*, was quoted as saying: "The HIV hypothesis ranks with the 'bad air' theory for malaria and the 'bacterial infection' theory of beriberi and pellagra [caused by nutritional deficiencies]. It is a hoax that became a scam" (April 3, 1994). See http://www.virusmyth.net/aids/controversy.htm.

57. Gilles Deleuze and Félix Guattari, *A Thousand Plateaus: Capitalism and Schizophrenia*, trans. Brian Massumi (Minneapolis: University of Minnesota Press, 1987), 10.

58. Mark Hansen, "Internal Resonance, or Three Steps Towards a Non-Viral Becoming," in *Culture Machine 3, Virologies: Culture and Contamination*, under the direction of Dave Boothroyd and Diane Morgan, (2001), online at http://culturemachine.tees.ac.uk/Cmach/Backissues/j003/Articles/hansen.htm.

59. Although four years later, it is has become less so. A French team has shown that some of the biggest viruses can be infected by smaller viruses, thus giving yet another sense to the hypervirus, and helping consider viruses alive. Cf. Didier Raoult et al., "The Virophage as a Unique Parasite of the Giant Mimivirus," *Nature* 455 (September 4, 2008): 100–104.

60. Lynn Margulis and Dorion Sagan, *Microcosmos* (New York: Summit Books, 1986), *What Is Life?* (New York: Simon and Schuster, 1995), *What Is Sex?* (New York: Simon and Schuster, 1997), *Slanted Truths* (New York: Springer-Verlag, 1997), and *Acquiring Genomes* (New York: Basic Books, 2002).

61. Hansen, "Internal Resonance."

62. Luciana Parisi, *Abstract Sex: Philosophy, Bio-Technology and the Mutations of Desire* (London: Continuum, 2004), 15 and 23.

63. David Baltimore, "Our Genome Unveiled," *Nature* 409 (February 15, 2001): 814–16.

64. Bruno Latour, "Portrait d'un biologiste en capitaliste sauvage," in *La clef de Berlin* (Paris: La Découverte, 1993), 100–129.

65. See Luc Boltanski and Ève Chiapello, *Le nouvel esprit du capitalisme* (Paris: Gallimard, 1999), 749.

66. Keith Ansell Pearson, *Viroid Life: Perspectives on Nietzsche and the Transhuman Condition* (London: Routledge, 1997), 150.

67. Steven Shaviro, "Two Lessons from Burroughs," in *Posthuman Bodies*, ed. Judith Halberstam and Ira Livingston (Bloomington: Indiana University Press, 1995), 38.

68. Ibid., 47.

69. Steven Shaviro, "William Burroughs," in *Doom Patrol*, chapter 10, online at http://www.dhalgren.com/Doom/ch10.html.

70. Louis Pawels, "Le monôme des zombis," *Le Figaro Magazine*, December 6, 1986.

71. Although Derrida consciously avoided the AIDS metaphor, critics were prompt to make the connection: "Derrida's most striking claim is that 9-11 is the result of an autoimmune disorder . . . 9-11 was a double suicide of both attackers and their victims. We are suffering from a metaphysical AIDS" (Gregory Fried, "The Uses of Philosophy," *Village Voice*, quoted in "Derrida: Democracy after 9/11," Philosophy.com, February 23, 2005, online at http://www.sauer-thompson.com/archives/philosophy/002925.html).

72. Jacques Derrida: "Autoimmune conditions consist in the spontaneous suicide of the very defensive mechanism supposed to protect the organism from external aggression," quoted in Giovanna Borradori, *Philosophy in a Time of Terror: Dialogues with Jürgen Habermas and Jacques Derrida* (Chicago: Chicago University Press, 2003), 150.

73. Ibid., 154.

74. Ibid., 187–88. Derrida refers here to his "Faith and Knowledge: The Two Sources of 'Religion' at the Limits of Reason Alone," in Jacques Derrida and Gianni Vattimo, eds., *Religion* (Stanford, Calif.: Stanford University Press, 1998).

75. Sadie Plant and Nick Land, "Cyberpositive," in *Unnatural: Techno-Theory for a Contaminated Culture*, ed. Matthew Fuller (London: Underground, 1994), available online at http://www.sterneck.net/cyber/plant-land-cyber/index.php, accessed December 28, 2008.

76. Sontag, "AIDS and Its Metaphors," 105–8, first emphasis in the original, second and third mine.

77. Borradori, *Philosophy in a Time of Terror*, 95.

78. Ibid.

79. Jean Baudrillard, "L'esprit du terrorisme," *Le Monde*, November 2, 2001, revised translation based on Rachel Bloul's translation, available on the Web page of the European Graduate School, accessed March 25, 2005, http://www.egs.edu/faculty/baudrillard/baudrillard-the-spirit-of-terrorism.html, accessed March 25, 2005.

80. Ibid.

81. William Burroughs, "The Priest They Called Him," from *Exterminator!* (1960), spoken word released on CD in 1993 with Kurt Cobain and Nirvana.

82. See Bukatman, *Terminal Identity*, 74–78, for an analysis of this figure.

83. The Unabomber was arrested in April 1996 and brought to court in 1997. "The capture of Ted Kaczynski has led to a 'mediamorphosis' of the suspected Unabomber—from chilling criminal to pop cultural icon" (Mark Dery, "Wild Nature," *21C* 4[96]: 47).

84. Although he planted one of his devices in an American Airlines flight.

85. Alston Chase reminds us that "during the seventeen years of the Unabomber's bombing spree, 1978–1995, America experienced more than 388,000 murders and over 22,000 bombings that killed an additional 386 people and injured 3,644 others" (*A Mind for Murder: The Education of the Unabomber and the Origins of Modern Terrorism* [New York: Norton, 2003], 25).

86. Ibid., 342.

87. Ibid., with picture of the gun in the pages inserted between pages 224 and 225.

88. Ibid., 226.

89. Letter sent to the *New York Times*, April 24, 1995, http://www.unabombertrial.com/manifesto/nytletter.html.

90. Chase, *A Mind for Murder*, 323. Chase adds that "the word 'power' appears in the manifesto 193 times, 'system' 210 times, and 'psychology' and 'psychologists' more than 65 times."

91. Ibid., 294.

92. "Alston Chase: Robert Birnbaum talks with the author of Harvard and the Unabomber" (posted April 28, 2003 on identitytheory.com), http://www.identitytheory.com/interviews/birnbaum102.html, accessed April 9, 2005.

93. Chase, *A Mind for Murder*, 232.

94. Paul McHugh, "The Making of a Killer" (review of "Harvard and the Unabomber: The Education of an American Terrorist," *First Things* 137 [November 2003]: 58–63), http://www.firstthings.com/ftissues/ft0311/reviews/mchugh.html, accessed April 9, 2005.

95. On this point, see Dr. Sally C. Johnson's *Psychiatric Competency Report* for the Unabomber trial (September 11, 1998), *Sacramento Bee*, Unabomber court documents homepage, http://www.unabombertrial.com/documents/psych_report1.html, accessed April 9, 2005.

96. Walter Gilbert, a leading researcher in recombinant DNA research, founded Biogen in 1978 (see Wikipedia entry "Biogen," http://en.wikipedia.org/wiki/Biogen_Idec, accessed August 7, 2007).

97. Cf. Michael Speaks, "Deux histoires pour l'avant-garde," http://www.archilab.org/public/2000/catalog/speaksfr.htm, my translation.

98. Fredric Jameson, "Future City," *New Left Review* 21 (May–June 2003), http://www.newleftreview.net/NLR25503.shtml, in a passage curiously omitted from the

reprint of the text in *Considering Rem Koolhaas and the Office for Metropolitan Architecture* (Rotterdam: Nai Publishers, 2003), 120–22.

99. Richard Meier, "Royal Gold Medal Address, Royal Institute of British Architects," in Rem Koolhaas and Bruce Mau, *S,M,L,XL*, ed. Jennifer Sigler (New York: The Monacelli Press, 1995), 1048 and 1052.

100. Koolhaas and Mau, *S, M, L, XL*, 960. Compare to "Other deteriorations, other exuberances followed at *Clarisse.* Populations and values have changed several times; the name says the same, and the most difficult objects to break . . . Perhaps Clarisse was never more than a mess of vestiges, chipped, sundry and unused . . . the world is covered by a single Trude that neither begins nor ends; all that changes is the airport's name" (Italo Calvino, *Les villes invisibles* [Paris: Seuil, 1972; 1996 for the pocket edition, collection Points 273, trans. J. Thibeaudeau], 126 for Clarisse and 149 for Trude in this edition).

101. This reference is tacit, but two of the three entries of the dictionary of *S, M, L, XL* attributed to Calvino ("inferno" and "exceptions") come from *Les villes invisibles.*

102. Rem Koolhaas, "The Generic City" (1994), in Koolhaas and Mau, *S, M, L, XL*, 1248. This essay concludes *S, M, L, XL*, by its date, but also because it handles architectural reflection on the grandest scale permitted: XXXL = global?

103. My translation. But to add at the outset: "The image of the 'megalopolis,' the continuous, uniform city also dominates my book. But there are already so many books that prophecy catastrophes and apocalypses that it would be redundant to write another, and that is certainly not in my nature" (Calvino, *Les villes invisibles*, vi, my translation).

104. Koolhaas, *S, M, L, XL*, 796 ; Calvino, *Les villes invisibles*, 189.

105. Calvino, *Les villes invisibles*, vi–vii.

106. Ibid., 84–85; Koolhaas and Mau, *S,M, L, XL*, 380.

107. Italo Calvino, "The Adventure of a Photographer," in *Difficult Loves* (London: Picador, 1985), cited in Koolhaas and Mau, *S, M, L, XL*, 1024.

108. Guy Debord, *La société du spectacle* (Paris: Gallimard, 1992 [originally published by Buchet-Chastel, Paris, 1967]), 9, my translation.

109. Koolhaas, "The Generic City," 1264.

110. "Junkspace" appeared for the first time in the *Harvard Design School Guide to Shopping* (Project on the City 2, ed. Chuihua Judy Chung, Jeffrey Inaba, Rem Koolhaas, Sze Tsung Leong, and Taschen, 2001), 408–21, before being taken up in the magazine *October* 100 (2002):"Obsolescence," 175–90. With this passage, I provide a (short, but I hope representative) list of excerpts.

111. Rem Koolhaas, "Skyscraper: A Typology of Public and Private," in *The State of Architecture at the Beginning of the 21st Century*, ed. Bernard Tschumi and Irene Cheng (New York: The Monacelli Press, 2003), 74.

112. Cf. Eric Schlosser, *Fast Food Nation* (New York: HarperCollins, 2002).

113. Walter Benjamin, *The Arcades Project*, ed. Rolf Tiedemann, trans. Howard Eiland and Kevin McLaughlin (Cambridge: Belknap Press of Harvard University Press, 1999), 3.

114. Jameson, "Future City."

115. "Traditionally, manufactured humans are either women or servants . . . There are thus two primary variations on the main theme: the artificial woman, or, more accurately, the artificial bride, and the artificial menials" (Michael Andermatt, "Artificial Life and Romantic Brides," in *Romantic Prose Fiction*, a volume in the ICLA (International Comparative Literature Association) Comparative Literary History Series, ed.: Gerald Gillespie (Stanford), Manfred Engel (Hagen), Bernard Dieterle (Technische Universität Berlin), http://homepage.sunrise.ch/mysunrise/mandermatt/publikation6.html.

116. http://www.artroca.com/art_folder/eve2.html.

117. Critical Art Ensemble, *Cult of the New Eve Position Paper*, http://www.critical-art.net/biotech/cone, accessed February 20, 2006.

118. But fortunately, as I revise this manuscript, Steve Kurz, one of the members of the Critical Art Ensemble, is now free.

119. "Microvenus—Art Form Using Genetic Sequences and Binary Code," *Art Journal* 55:1 (Spring 1996) 70–75, http://www.findarticles.com/p/articles/mi_m0425/is_n1_v55/ai_18299596.

120. Adam Zaretsky, "The Mutagenic Arts," *CIAC's Electronic Magazine* 23 (fall 2005), http://www.ciac.ca/magazine/archives/no_23/en/index.html.

121. http://www.ekac.org/transgenicindex.html.

122. Ibid.

123. "Eduardo Kac's Genesis: Biotechnology between the Verbal, the Visual, the Auditory, and the Tactile," Installation at the Julia Friedman Gallery, Chicago, Reviewed by Simone Osthoff, assistant professor of art criticism, School of Visual Arts, Penn State University. Originally published in *Leonardo Digital Reviews*, October 2001, http://mitpress2.mit.edu/ejournals/Leonardo/reviews/oct2001/ex_GENESIS_osthoff.html, accessed February 20, 2006, available at http://www.ekac.org/osthoffldr.htm.

124. About this tragic phenomenon, one will consult with great advantage Thomas Frank's *The Conquest of Cool* (Chicago: University of Chicago Press, 1997), and Joseph Heath and Andrew Potter, *The Rebel Sell* (Toronto: HarperCollins, 2004). Or, see Frédéric Beigbeder, *14.99€* (Paris: Grasset, 2001).

125. Joe Davis, *Microvenus*.

126. http://www.ekac.org/move36.html.

127. Zaretsky, "The Mutagenic Arts."

128. Slavoj Žižek, *Organs without Bodies: On Deleuze and Consequences* (London: Routledge, 2004), 121.

129. Ionat Zurr and Oron Catts, "Artistic Life Forms That Would Never Survive Darwinian Evolution: Growing Semi-Living Entities," http://www.tca.uwa.edu.au/atGlance/pubMainFrames.html.

130. http://www.tca.uwa.edu.au/ars/text.html.

131. For instance, about their collaboration with Stelarc, they note that their work involves "the actual and suggestive disfigurement of the human body—the detached

organ which is easily recognizable as human—a somewhat playful reverse reference to Artaud's body without organs was in our case an organ with no body; or rather an organ with a technological body" (Oron Catts and Ionat Zurr, "The Art of the Semi-Living and Partial Life: Extra Ear–1/4 Scale," available on their Web site: http://www.tca.uwa.edu.au/publication/TheArtoftheSemi-LivingandPartialLife.pdf).

132. Ibid.

133. Žižek, *Organs without Bodies,* 120–21.

134. Gershom Scholem, "The Golem of Prague and the Golem of Rehovot," in *The Messianic Idea in Judaism and Other Essays in Jewish Spirituality* (New York: Schocken, 1971), 338.

135. Sonya Rapoport, "Redeeming the Gene, Molding the Golem, Folding the Protein," online at http://users.lmi.net/sonyarap/redeeming/index.html.

136. Gershom Scholem, "Isaac Luria and His School," in *Major Trends in Jewish Mysticism* (New York: Schocken, 1946 [1995]), 281.

De-Coda

1. Richard Doyle, "LSDNA: Rhetoric, Consciousness Expansion, and the Emergence of Biotechnology," *Philosophy and Rhetoric* 35(2) (2002): 153–74.

2. Ibid.

3. Alun Rees, "Nobel Prize Genius Crick Was High on LSD When He Discovered the Secret of Life," *Mail on Sunday* (London), August 8, 2004.

4. Guy Debord, *La société du spectacle* (Paris: Gallimard, 1992 [1967]), 53. This thesis, numbered 57, applied originally to "third-world revolutionaries," in Debord's mind. What would have been his despair to realize that it applies equally well to the "revolution" that he, among others, started.

5. Jean Baudrillard, *Forget Foucault* (New York: Semiotext[e], 2006 [1977]), 46.

6. Pascal, *Pensées* 347.

7. William Gibson, "The Winter Market," in *Burning Chrome* (New York: Arbor House, 1986).

8. Korzybski's insights were adopted by the psychotherapist Albert Ellis, William S. Burroughs (who took his seminar in the late 1930s), and A. E. van Vogt (who in turn greatly influenced PKD); his problematization of cause–effect language and thinking into General Semantics (resting on non-Aristotelian, non-Newtonian, and non-Euclidian axiomatics) were borrowed by L. Ron Hubbard in his science fiction and other work (!).

9. In spite of William Gibson's claim that Philip K. Dick "had almost no influence" on him, I consider that PKD is the most crucial cyberfiction writer. Anyway, Gibson's writings, in his own words, contradict his claim. See Larry McCaffery's interview with William Gibson (August 1986) in *Storming the Reality Studio: A Casebook of Cyberpunk and Postmodern Science Fiction* (Durham, N.C.: Duke University Press, 1991), 277, and Darren Wershler-Henry, "Queen Victoria's Personal Spook, Psychic Legbreakers, Snakes and Catfood: An Interview with William Gibson and Tom Maddox,"

Virus 23(0) (Fall 1989): 28–36. In both cases, Gibson talks about the presence of junk in his writing, but the second time, he uses PKD's word for junk, *kipple.* No influence, right, Dr. Gibson?

10. Horace Freeland Judson, *The Eighth Day of Creation* (Cold Spring Harbor, N.Y.: Cold Spring Harbor Laboratory Press, 1996), 333–34.

11. See Gregory Stock, *Redesigning Humans: Our Inevitable Genetic Future* (Boston: Houghton Mifflin, 2002).

12. James Morrow's Godhead Trilogy (*Towing Jehovah,* 1994; *Blameless in Abaddon,* 1996; *The Eternal Footman,* 1999).

13. ENCODE Project Consortium, "Identification and Analysis of Functional Elements in 1% of the Human Genome by the ENCODE Pilot Project," *Nature* 447 (June 14, 2007): 799–816. The pilot project of the ENCyclopedia Of DNA Elements (ENCODE) project consortium is a systematic study of a targeted 29,998 kilobases (roughly 1 percent), aiming at providing "a more biologically informative representation of the human genome by using high-throughput methods to identify and catalogue elements encoded" (799). It is thus the acme of today's research on hyperreal DNA (see chapter 2). Its first results, published in the June 14, 2007, issue of *Nature* and in no fewer than twenty-eight companion papers that same month in *Genome Research,* created quite a stir: they found that DNA was much more "pervasively transcribed" than was previously thought, and it attracted a lot of attention. Francis S. Collins, the director of the National Human Genome Research Institute (NHGRI, a part of the U.S. National Institutes of Health), which organizes the Consortium, was prompt to react to the news: "This impressive effort has uncovered many exciting surprises and blazed the way for future efforts to explore the functional landscape of the entire human genome," he said, before adding that "the scientific community will need to rethink some long-held views about what genes are and what they do, as well as how the genome's functional elements have evolved" (U.S. Department of Health and Human Services, "New Findings Challenge Established Views on Human Genome," *NIH News,* Bethesda, June 13, 2007, online at http://www.genome.gov/25521554, accessed June 13, 2007). Others were even more dramatic in their comments: a science reporter wrote that "this tidy picture of how genes work [the picture provided by the central dogma of genetics] has been muddied by a mammoth investigation of human DNA" (Andy Coghlan, "'Junk' DNA Makes Compulsive Reading," *New Scientist* 2608 [June 13, 2007]: 20). In this article, he gave the last word to John Greally, a scientist with the Albert Einstein College of Medicine in New York City, who declared: "it would now take a very brave person to call non-coding DNA junk." The challenging result of the ENCODE project is also a quantitative result: "93% of bases were represented in a primary transcript identified by at least two independent observations (but potentially using the same technology); this figure is reduced to 74% in the case of primary transcripts detected by at least two different technologies" (803). Pervasively transcribed, all right! Noncoding RNA, that is, the processed junk DNA sequences that are transcribed but not translated (into proteins), might play a part of crucial importance in

this semiosis, making them the interpretants of choice. This is the main *really* "challenging result" of the ENCODE study—challenging to the dogmatic model, that is, since it does not "challenge" at all the ideas of Roy Britten or John Mattick, for instance (see chapter 2). Previous works on the importance of RNA further allow us to make the following hypotheses: (1) the bootstrapping of the living semiosis was taken over by RNA from the start, making of ncRNA a true oxymoron, and (2) rather than considering one or a set of RNAs as a bootstrap program, we should consider the possibility of the existence of numerous coevolving ncRNAs emerging as a bootstrap network.

14. See Mark B. Gerstein et al., "What Is a Gene, Post-ENCODE? History and Updated Definition," *Genome Research* 17 (2007): 669–81, and Thomas R. Gingeras, "Origin of Phenotypes: Genes and Transcripts," *Genome Research* 17 (2007): 682–90, for two opposite takes on the question of the obsolescence of the gene. I side with Gingeras here: "while the concept of the gene has been helpful in defining the relationship of a portion of a genome to a phenotype," he writes in his abstract, "*this traditional term may not be as useful as it once was*" (my emphasis).

15. What is needed, in fact, is to move this notion of sign to the Peircian concept of *semiosis*, an endless sign process where the interpretant of a given sign can become the sign-vehicle of another sign. Peirce defined a sign as "anything which is so determined by something else, called its Object, and so determines an effect upon a person, which effect I call its interpretant, that the later is thereby mediately determined by the former" (in *Semiotics and Significs: The Correspondance between Charles S. Peirce and Victoria Lady Welby*, ed. C. S. Hardwick and J. Cook [Bloomington: Indiana University Press, 1977 (1908)], 80–81). The sign is also the triadic relation between object and interpretant via a signifying element, alternatively called "sign," "representamen," "representation," "ground," or, my choice here, "sign-vehicle." The idea is not new and can be found as early as 1979 in Thomas Sebeok's writings: "a full understanding of the dynamics of semiosis may in the last analysis turn out to be no less than the definition of life" (Thomas A. Sebeok, *The Sign and Its Masters* [Austin: University of Texas Press, 1979], 26). See also Kalevi Kull, "On Semiosis, Umwelt, and Semiosphere," *Semiotica* 120(34) (1998): 299–310. What is new, however, is the possible convergence of a better understanding of nucleic acids (DNA and all classes of RNA) and a semiotic understanding of life. Of particular interest here is the idea that so-called noncoding RNA (ncRNA) could participate in a "metalanguage" (implying some sort of metacode) in the semiosis of the living. See Yair Neuman, "A Point for Thought: Does the Genetic System Include a Meta-Language?" in *Reviving the Living: Meaning Making in Living Systems* (Amsterdam: Elsevier, 2008), 55–74. The important point here is the non-univocality, or even the equivocality, of interpretation in the living semiosis. Moreover, one more consequence of this idea might imply the inclusion in this revised picture of living onto- and morphogenesis of *nongenetic sign vehicles*. The alluded RNA metacode might thus also concern proteins (e.g., histones) or chemicals (e.g., methyl groups) in this overall picture of semiosis, hereby encompassing what is usually referred to as "epigenetic processes." One more step would even add to this list

of potential signs some environmental factors (e.g., stress indicators) interpreted along different genetic and epigenetic processes such as transposition, for instance (as in the work of Barbara McClintock). If proven correct, or even partially correct, the abovementioned intuitions of a general semiosis of the living might lead to a revised model of the nature and functions of nucleic acids: *an ecological model.* After a good conversation with Kalevi Kull, I should add that the very schematic picture of biosemiosis presented here should not be taken to mean that there should exist one (and one only) network involving molecules of all kinds and environment factors in living onto- and morphogenesis. The ecological model should instead be based on the assumption that there many such networks, sometimes isolated from one another, sometimes coordinated; sometimes cooperating, sometimes competing. The results of these distributed processes, messy though they are, might very well be life itself; contingent, fragile, paradoxical, and immensely complex life. Personal communication at the eighth gathering in Biosemiotics, University of the Aegean, Syros Island, Greece, June 26, 2008.

16. Giorgio Agamben, "Bartelby, or On Contingency," in *Potentialities: Collected Essays in Philosophy,* ed. and trans. Daniel Heller-Roazen (Stanford, Calif.: Stanford University Press, 1999), 266.

17. Ibid.

18. Giorgio Agamben, "Tradition of the Immemorial,"in *Potentialities,* 106.

Glossary

1. Dov. S. Greenbaum, "Junk?" Genomics & Bioinformatics MBB 452a, Yale University, http://bioinfo.mbb.yale.edu/mbb452a/projects/Dov-S-Greenbaum.html.

2. Human Genome Nomenclature Committee, "Draft Guidelines for Human Gene Nomenclature" (2000), http://www.gene.ucl.ac.uk/nomenclature/guidelines/draft_2001.html.

3. Richard A. Pizzi, "Genetic Ciphering," *Modern Drug Discovery* 4(3) (March 2001): 65–66, http://pubs.acs.org/subscribe/journals/mdd/v04/i03/html/03timeline.html.

4. National Human Genome Growth Institute, "Summary of the Initial Sequencing and Analysis of the Human Genome," press release, February 11, 2001, http://www.nhgri.nih.gov/NEWS/summary_of_sequence.html.

Index

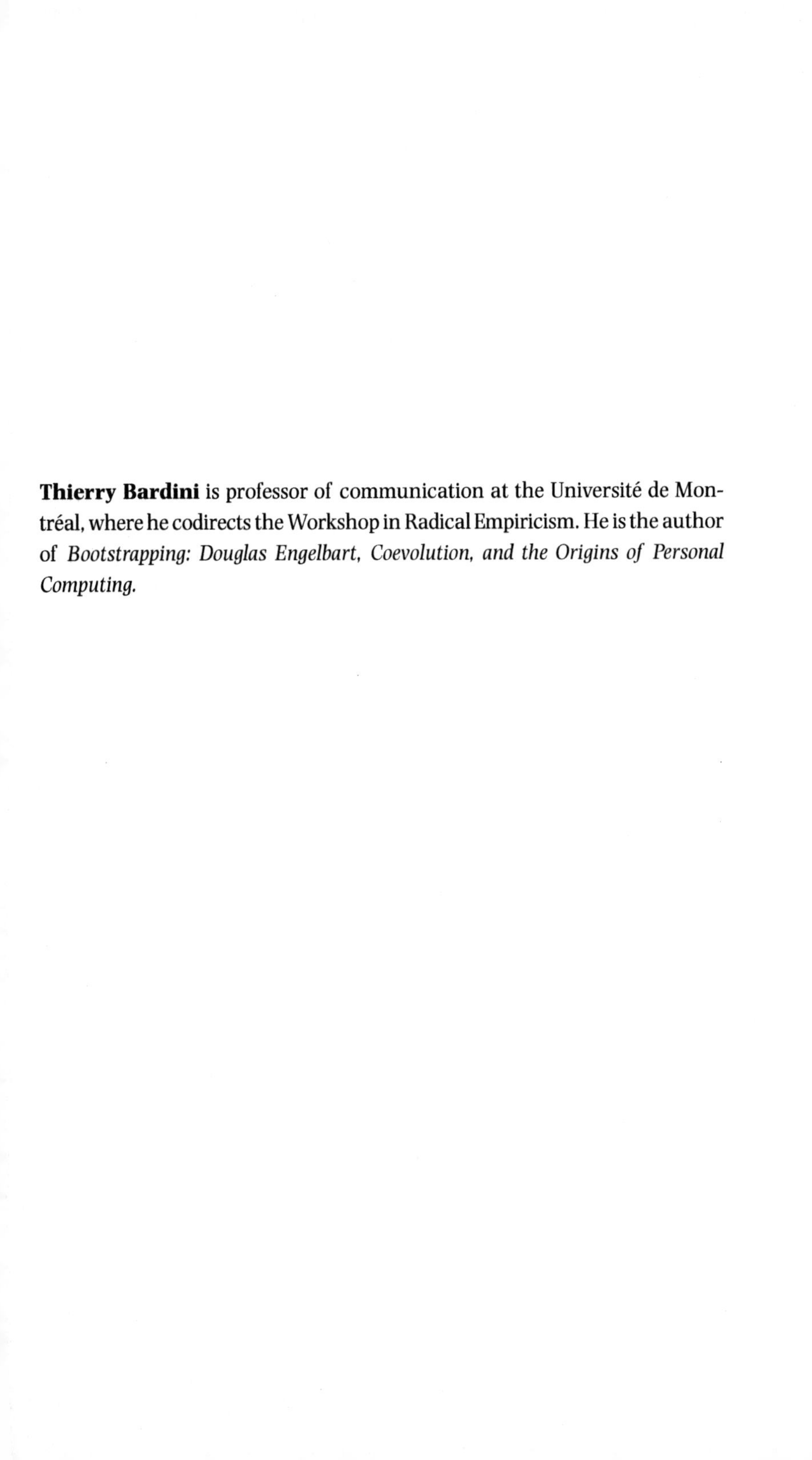

Thierry Bardini is professor of communication at the Université de Montréal, where he codirects the Workshop in Radical Empiricism. He is the author of *Bootstrapping: Douglas Engelbart, Coevolution, and the Origins of Personal Computing.*